Geocomputation

Geocomputation
a Primer

Edited by

PAUL A LONGLEY
SUE M BROOKS
RACHAEL MCDONNELL
BILL MACMILLAN

JOHN WILEY & SONS

Chichester • New York • Weinheim • Brisbane • Singapore • Toronto

Other Wiley Editorial Offices

John Wiley & Sons, Inc., 605 Third Avenue,
New York, NY 10158-0012, USA

WILEY-VCH Verlag GmbH, Pappelallee 3,
D-69469 Weinheim, Germany

Jacaranda Wiley Ltd, 33 Park Road, Milton,
Queensland 4064, Australia

John Wiley & Sons (Asia) Pte Ltd, 2 Clementi Loop #02-01,
Jin Xing Distripark, Singapore 129809

John Wiley & Sons (Canada) Ltd, 22 Worcester Road,
Rexdale, Ontario M9W 1L1, Canada

Library of Congress Cataloging-in-Publication Data

Geocomputation : a primer / edited by Paul A. Longley . . . [et al.].
    p. cm.
   Includes bibliographical references and index.
   ISBN 0-471-98575-9. — ISBN 0-471-98576-7
   1. Geographic information systems. 2. Information storage and
retrieval systems—Geography. I. Longley, Paul.
  G70.G442 1998
  910'.285—dc21   98–27225
                                                   CIP

British Library Cataloguing in Publication Data

A catalogue record for this book is available from the British Library

ISBN 0-471-98575-9 cloth
     0-471-98576-7 paper

Typeset in 10/12pt Times by Mayhew Typesetting, Rhayader, Powys
Printed and bound in Great Britain by Bookcraft (Bath) Ltd
This book is printed on acid-free paper responsibly manufactured from sustainable forestry,
in which at least two trees are planted for each one used for paper production.

Contents

List of Contributors

Malcolm G Anderson, School of Geographical Sciences, University of Bristol, University Road, Bristol BS8 1SS, UK.
(Research interests: slope stability modelling, including soil, climate and vegetation change, the effects on hydrological processes and slope stability. Telephone: +44-117-9287-871; fax: +44-117-9287-878; email: M.G.Anderson@bris.ac.uk)

Luc Anselin, Bruton Center for Development Studies and School of Social Sciences, University of Texas at Dallas, Richardson, TX 75083-0688, USA.
(Research interests: spatial econometrics and spatial statistics, GIS and spatial analysis, statistical computing, regional economic and demographic analysis. Telephone: +1-972-883-2088; fax: +1-972-883-2735; email: lanselin@utdallas.edu)

Peter M Atkinson, Department of Geography, University of Southampton, Southampton SO17 1BJ, UK.
(Research interests: geostatistics, remote sensing, GIS. Telephone: +44-1703-594-617; fax: +44-1703-593-295; email: pma@soton.ac.uk)

Michael Batty, Centre for Advanced Spatial Analysis, University College London, 1–19 Torrington Place, London WC1E 6BT, UK.
(Research interests: cellular automata, fractal geometry, VR, GIS, urban cyberspace. Telephone: +44-171-391-1782; fax: +44-171-813-2843; email: mbatty@geog.ucl.ac.uk)

Sue M Brooks, School of Geographical Sciences, University of Bristol, University Road, Bristol BS8 1SS, UK.
(Research interests: slope stability modelling, including soil, climate and vegetation change, the effects on hydrological processes and slope stability. Telephone: +44-117-9289-109; fax: +44-117-9287-878; email: Susan.Brooks@bris.ac.uk)

Peter A Burrough, Utrecht Centre for Environment and Landscape Dynamics, Faculty of Geographical Sciences, Utrecht University, Post Box 80.115, 3508 TC Utrecht, the Netherlands.

(Research interests: spatial analysis, geostatistics, environmental modelling with GIS. Telephone: +31-30-2532-749; fax: +31-30-2540-604; email: p.burrough@frw. ruu.nl; home page: http://www.frw.ruu.nl/pcraster.html)

Keith C Clarke, Department of Geography and National Center for Geographic Information and Analysis, University of California, Santa Barbara, CA 93106-4060, USA.
(Research interests: cartography, geographic information science. Telephone: +1-805-893-7961; fax: +1-805-893-3146; email: kclarke@geog.ucsb.edu; home page: http://www.geog.ucsb.edu/~kclarke)

Andrew D Cliff, Department of Geography, University of Cambridge, Downing Place, Cambridge CB2 3EN, UK.
(Research interests: spatial diffusion, epidermiology, spatial statistics. Telephone: +44-1223-333-399; fax: +44-1223-333-392; email: adc2@hermes.cam.ac.uk)

Helen Couclelis, Department of Geography and National Center for Geographic Information and Analysis, University of California, Santa Barbara, CA 93106-4060, USA.
(Research interests: urban and regional modelling and planning, spatial cognition, geographic information science. Telephone: +1-805-893-2196; fax: +1-805-893-3146; email: cook@geog.ucsb.edu)

Paul J Curran, Department of Geography, University of Southampton, Southampton SO17 1BJ, UK.
(Research interests: remote sensing at local to global scales, ecosystem modelling. Telephone: +44-1703-592-259; fax: +44-1703-593-295; email: ggcurran@soton. ac.uk)

Martin Dodge, Centre for Advanced Spatial Analysis, University College London, 1–19 Torrington Place, London WC1E 6BT, UK.
(Research interests: cyber-geography, GIS, VR, mapping cyberspace. Telephone: +44-171-391-1782; fax: +44-171-813-2843; email: m.dodge@ucl.ac.uk)

Simon Doyle, Centre for Advanced Spatial Analysis, University College London, 1–19 Torrington Place, London WC1E 6BT, UK.
(Research interests: GIS in archaeology, VR, data analysis, spatial analysis. Telephone: +44-171-391-1782; fax: +44-171-813-2843; email: s.doyle-walsh@ucl. ac.uk)

Giles M Foody, Department of Geography, University of Southampton, Southampton SO17 1BJ, UK.
(Research interests: environmental remote sensing, mapping, quantitative methods. Telephone: +44-1703-595-493; fax: +44-1703-593-295; email: gmf@soton.ac.uk)

Michael F Goodchild, Department of Geography and National Center for Geographic Information and Analysis, University of California, Santa Barbara, CA 93106-4060, USA.
(Research interests: geographic information systems, spatial analysis, accuracy and uncertainty in geographic data, spatial data libraries, geospatial interoperability. Telephone: +1-805-893-8049; fax: +1-805-893-7095; email: good@ncgia.ucsb.edu)

Peter Haggett, School of Geographical Sciences, University of Bristol, University Road, Bristol BS8 1SS, UK.
(Research interests: spatial diffusion, epidermiology, locational analysis. Telephone: +44-117-9289-171; fax: +44-117-9287-878; email: P.Haggett@bris.ac.uk)

Paul A Longley, School of Geographical Sciences, University of Bristol, University Road, Bristol BS8 1SS, UK.
(Research interests: GIS, spatial analysis, data integration, RS–GIS integration, survey research practice. Telephone: +44-117-9287-509; fax: +44-117-9287-878; email: Paul.Longley@bris.ac.uk)

Bill Macmillan, School of Geography, University of Oxford, Mansfield Road, Oxford OX1 3TB, UK.
(Research interests: optimisation, conceptual issues in quantitative geography and regional science. Telephone: +44-1865-271-942; fax: +44-1865-271-929; email: bill.macmillan@geography.oxford.ac.uk)

Rachael McDonnell, School of Geographical Sciences, University of Bristol, University Road, Bristol BS8 1SS, UK.
(Research interests: hydrology, water quality, environmental modelling, arid zones. Telephone: +44-117-9289-000; fax: +44-117-9287-878; email: Rachael_McDonnell@ compuserve.com)

Edward J Milton, Department of Geography, University of Southampton, Southampton SO17 1BJ, UK.
(Research interests: field spectrometry, quantitative remote sensing. Telephone: +44-1703-593-260; fax: +44-1703-593-295; email: ejm@soton.ac.uk)

Stan Openshaw, Centre for Computational Geography, School of Geography, University of Leeds, Leeds LS2 9JT, UK.
(Research interests: the three core geocomputational tool-sets – artificial intelligence, GIS, high-performance computing applications in geography. Telephone: +44-113-2333-320; fax: +44-113-233-3308; email: stan@geography.leeds.ac.uk)

Andy Smith, Centre for Advanced Spatial Analysis, University College London, 1–19 Torrington Place, London WC1E 6BT, UK.
(Research interests: participation in planning, VR, networks, CAD. Telephone: +44-171-391-1782; fax: +44-171-813-2843; email: asmith@geog.ucl.ac.uk)

Preface

This is a conference volume with a difference! The field of geocomputation is a new and fast-developing one, and in September 1998 the Third International Conference on the subject is being held at the School of Geographical Sciences, University of Bristol, UK. The local organising committee have been very fortunate in securing the participation of the key international researchers in the field, and the format of the conference will be structured around their keynote presentations. These same individuals have produced written contributions to exacting and detailed style and content guidelines, each of which blends state-of-the-art research with advanced pedagogy: together they comprise a clear, comprehensive and thoroughly up-to-date statement of what is new and exciting about geocomputation today.

This book will be made available to all conference participants, but we see it as of wider import than a guide to a conference. Today there are more and more postgraduate and advanced undergraduate courses in the areas discussed in this book, and our hope and expectation is that the contributions to this volume will provide students on these courses with a catalyst to greater understanding of what geocomputation is and what it entails. Just as it will, we hope, structure a very successful international conference, so we hope it will provide a more enduring companion to the study and practice of spatial computing.

Paul Longley
Sue Brooks
Rachael McDonnell
Bill Macmillan
Bristol, May 1998

Acknowledgements

Helen Couclelis gratefully acknowledges the support provided for her chapter by the journal **Environment and Planning B: Planning and Design** and Pion Ltd.

Michael Goodchild acknowledges the support of the US National Center for Geographic Information and Analysis and the Alexandria Digital Library, which are supported by the US National Science Foundation.

Part of the research reported in **Luc Anselin**'s chapter was supported by Grant SBR-9410612 from the US National Science Foundation. Shuming Bao provided insightful comments and made the beta version of the S+ArcView interface available. Arc/Info, ArcView and Avenue are trademarks of ESRI, S-Plus, S+Gislink, S+ArcView, and S+API are trademarks of MathSoft Inc., XploRe is a trademark of W Härdle.

Stan Openshaw's research is supported by ESRC grant R237260. He is grateful to Ian Turton for assistance, and would also like to acknowledge the use of 1991 Census data and boundary data purchased by ESRC/JISC. All data and maps hold Crown Copyright.

Peter Burrough wishes to thank Cess Wesseling, Willem van Deursen, Derk Jan Karssenberg and Ad de Roo for providing the tools, the data, and many of the ideas for his chapter.

Andrew Cliff and Peter Haggett are grateful to the Wellcome Trust for a programme grant which has allowed them to explore here (and in greater detail elsewhere) the spatial diffusion issues considered in their chapter.

While every effort has been made to trace the owners of copyright material, we offer our apologies to any copyright holders whose rights we may have unwittingly infringed.

Part One
ON GEOCOMPUTATION

1

Foundations

Paul A Longley

1.1 Introduction

This is a book concerning all that is new about 'geocomputation', yet its subtitle – 'a primer' – indicates that it also seeks to set out the foundations and scope of this area from the same contemporary perspective. The catalyst to the recent adoption of the term 'geocomputation' to describe a range of research activities has undoubtedly been the application of computers to devise and depict digital representations of the Earth's surface. The environment for geocomputation is provided by geographical information systems (GIS: Longley et al 1998), yet what is distinctive about geocomputation is the creative and experimental use of GIS that it entails. The hallmarks of geocomputation are those of research-led applications which emphasise process over form, dynamics over statics, and interaction over passive response.

Like all more mainstream GIS applications, geocomputational representations are digital 'models', simplifications of geographical reality, at various levels and to varying degrees of abstraction. This process of abstraction by definition discards information, whether the underlying purpose is to model spatial distributions (data models), flows (e.g. spatial interaction models), networks, hierarchies, or diffusion (as set out in the classic exposition of Haggett et al 1977). The fundamental technology of GIS has undoubtedly developed and matured in recent years, yet GIS remains a tool which can be used for a wide range of scientific and general ends. In itself GIS provides little guidance as to how best spatial problems might be solved. It is clear that big technical and epistemological questions remain about the ways in which selected facets of geographical reality are discarded or emphasised in the process of model-building. Geocomputation provides a framework within which those researching the development and application of GI technologies

Geocomputation: A Primer. Edited by Paul A Longley, Sue M Brooks, Rachael McDonnell and Bill Macmillan.
© 1998 John Wiley & Sons Ltd.

are seeking to address and resolve many of these important questions. The titles of the different parts of this book suggest the range of ways in which this is being achieved: through developing improved, tailored data models of spatial distributions, through visualisation and linkage, and through explicit process modelling. The spirit of geocomputation is fundamentally about matching technology with environment, process with data model, geometry and configuration with application, analysis with local context, and philosophy of science with practice.

As Couclelis (this volume) and Macmillan (this volume) are at pains to point out, there are a number of ways in which geocomputation is significantly different from mainstream GIS. Any survey of research papers about 'geocomputation' will reveal that many researchers in the field do not use standard proprietary GIS technology, and choose instead to write their own programs in order to put methods into practice. Others prefer to couple GIS modules to their own or other specialised software in a variety of ways (Anselin, this volume), and with the current trend towards software interoperability the distinctions between geocomputation and GIS practice will likely blur still further. For the time being, however, GIS is one of many tools in geocomputation, and should be viewed essentially as part of the rapid secular developments in general computing.

In all these respects, geocomputation is avowedly empiricist, and much has been written in recent years about the extent to which empiricist approaches are embedded in their technological context (e.g. Macmillan 1989). Over the last 30 years, the role of the computer has developed from that of scientific tool used for specific ('analysis') stages of the research process to a wider, more integrated and (arguably) more holistic role in data collection, exploration, transformation, visualisation, and even presentation and report-writing. Thus, as Goodchild and Longley (1998) have noted, the computer is no longer merely an adjunct to the research environment – today it *is* the research environment, and the contributors to this book variously describe the ways in which geocomputation may be used throughout the process of research design and analysis. Couclelis (this volume) sees this development as marking the transition from the computer as a contained 'equation-solving and data-handling device' to its becoming 'an integral component of the modelling of complex systems and processes', and echoes of this argument may be discerned in most of the other contributions.

This transformation has taken place in the wake of precipitous falls in the cost of computer power, informally codified as Moore's Law (named after Gordon Moore, co-founder of the microprocessor company Intel), which suggests that computer hardware performance doubles and price halves every 18 months (Longley et al 1998). Cheaper and cheaper computing power has had two immediate and profound implications for geocomputation. First, some of these benefits have been used to develop ever more sophisticated graphical user interfaces (GUIs), based on the now ubiquitous WIMP (windows, icons, mice, pointers) model and extended into the realm of 3-D and virtual reality using avatars (Batty et al, this volume). These are having the effect of making computation accessible (and apparently intelligible) to the many, rather than restricting it to the preserve of the few. Second, vastly enhanced processing power is extending the scope of computation to more and more of the many, who can now use computation for visualisation, exploration and experiment. These developments have in turn stimulated change in the way that GIS software is marketed and developed – specifically through commercial off-the-shelf (COTS) GIS (Maguire 1998). Against this general background, the quite recent innovation of the Internet has triggered the fragmentation of previously monolithic GIS packages into reusable software modules. These have been scattered across the Internet, which has become a new important platform for GIS. At the same time, similar trends in the pricing of software and data have led to their pre-packaging into 'desk-top' systems (Elshaw Thrall and Thrall 1998). The carcass of GIS, its hardware, is also evolving from desk-based shells through laptops, palm-tops and pen-computers into a variety of portable and wearable forms. Clarke's (this volume) view is that this will soon permit the creation of digital representations in the field and 'on the fly' with roaming data abstractors and modellers. And finally, the data bottleneck, which until very recently so restricted the scope and richness of computational modelling, has burst. Rapid developments have taken place in the creation of 'framework' data, defined in a general sense by Smith and Rhind (1998) as 'a core set of standard variables which are commonly used or needed and which ideally are available over the whole domain of interest' such data include (for example) complete digital map coverage of the UK – although reappraisal and reform of the ways in which intellectual property rights are recouped may be necessary before they can truly be considered to be part of the digital data infrastructure of information economies. Elsewhere, however, developments in satellite imaging (Curran et al, this volume) and global positioning systems (GPS: Clarke, this volume) are leading to rapid proliferation of framework datasets across a wider range of domains than ever before, holding the prospect of genuinely 'data-rich' modelling for the first time. And more and more attribute data are becoming available in the socioeconomic realm (Birkin 1995), which may be exploited to devise richer 'digital personae' as components of models of human activity patterns if problems of respondent self-selectivity and bias can be successfully accommodated.

All of these rapid changes owe more to developments in technology than science, and this book sets out to clarify the extent to which the science and practice of geocomputation might be considered essentially 'reactive' to the rapidly developing technological setting (e.g. technology led), or whether it is more accurately portrayed in a more active sense as the quest to develop new science using new tools (see Batty and Longley 1996 for a similar discussion of the 'spatial analysis in GIS' vs. 'GIS in spatial analysis' debate; and Thrift 1981 for an earlier discussion of the analogous development of behavioural geography). This is much more than just an academic or moot point: for some, Openshaw's (this volume) new automated 'artistic science of statistical analysis' is over-egging a more mundane reality – namely that geocomputation is essentially reactive, a 'black box' technological response, which is diverting attention from more truly active and rigorous use of GIS tools to develop GI Science (Couclelis, this volume; Goodchild 1992). Technology empowers us with tools, yet conventional wisdom asserts that we need consciously and actively to use them in developing science without surrendering control to the machine. Do better models invariably require better theories, such as those espoused by Burrough (this volume) in the context of dynamic modelling? Or has the esteem in which we hold traditional deductive science diverted us from the equally valid quest for predictive success, which Openshaw (this volume) contends is demonstrably best attained through computationally intensive inductive generalisation? A middle path is suggested by Mandelbrot's (1986) observation that '[d]escription coming before theory is the normal pattern in science', and the view that visualisation of pattern suggests insight into generating process (Batty and Longley 1994). Visualisations and virtual realities can most certainly enhance our understanding of scientific problems, particularly if they are shared among many users who can each bring their own expertise and insights to collective problem-solving (Batty et al, this volume; Shiffer 1998). From ontology to prediction and practice, internal debate is likely to characterise the evolution of geocomputation although, as Couclelis (this volume) observes, there is no obvious single meta-theoretical framework for activity in prospect. While this may chime a chord with current postmodern fashions, it nevertheless provides the area with much looser anchorings than its antecedents in quantitative geography (Cliff and Haggett, this volume).

1.2   The Developing Remit of Geocomputation

These debates are common to wide areas of science and social science, yet what is distinctive about geocomputation is the ways in which new computational tools and methods are used to depict geography across the range of scales from the architectural to the global. The innovation of cheap, high-speed computing is allowing traditional problems of spatial association, dependence and autocorrelation to be viewed from a range of new perspectives. As the cover of this book suggests, geocomputation is also fundamentally about the depiction of spatial *process*, in stark contrast to the essentially static framework of much earlier work in geography and the other geosciences. Thus our cover implies a range of processes as diverse as desiccation, hydrological force and urban growth shown in a temporal context. Many established dynamic modelling approaches in geography are being recast in the data-rich, computer-intensive geocomputational environment, and it is no longer necessary to treat time as a linear and mechanistic trajectory. Indeed, in the spirit of Dawkins (1986) geocomputation is also being used to conceive of temporal change in a range of evolutionary and non-linear senses with respect to the evolution of natural (e.g. river deltas: Burrough, this volume) and artificial (e.g. city morphologies: Batty and Longley 1994) forms.

How has geocomputation developed? The wide usage of the term originates from Bob Abrahart's masterminding the highly successful 'First International Conference on Geocomputation' in Leeds (UK) in September 1996. However, earlier references to the term 'geocomputation' appear in some of the papers of Openshaw in the mid-1990s and, as his contribution to this volume makes clear, the term aptly describes the furrow that he has ploughed for some 20 years (e.g. Openshaw and Taylor 1979). For most readers, Openshaw's (this volume) geographical analysis and explanation machines epitomise the spirit and purpose of geocomputation, while for others (for reasons set out in detail by Couclelis, this volume) this remains a very particular view of science and generalisation which is viewed with some scepticism. Perusal of the programmes of the first two geo-computation conferences suggests that the constituency of geocomputation is drawn from a much wider range of perspectives than this and, as Couclelis points out, there is no universality of view as to precisely what the 'geocomputation paradigm' entails. Indeed some (e.g. Macmillan 1998) see strong elements of con-tinuity between geocomputation and the established traditions of quantitative geography – because it shares the same analytical philosophy and the same scien-tific methods that flow from it, and because established techniques (such as spatial interaction modelling and Markov Chains) have been successfully developed and extended in the GIS environment. Yet for others (e.g. Couclelis, this volume), geocomputation is viewed as fundamentally more uncoordinated in ontological and epistemological terms, if indeed it can be described as 'theoretical' at all.

How has this state of affairs arisen? The standard quantitative 'linear project design' (Goodchild and Longley 1998) – wherein problem conception leads to hypothesis formulation, a sample design, data collection, coding and analysis, and finally report-writing – presents a transparent, rational, informed and internally consistent approach to scientific reasoning which has underlain generations of research projects. Yet this ordered, rational perspective sits uncomfortably in an emergent digital world characterised by the uncoordinated proliferation of digital datasets (many of which are collected by unorthodox or even profoundly unscien-tific means), the scattering of data and software across the World Wide Web and the (relative, at least) trend away from publicly funded data acquisition to private data warehousing. Together, these trends may be increasing the richness and diversity of digital representations, but the analytical approaches that they foster are becoming increasingly at odds with the validity of the scientific 'truths' that they generate. Peter Taylor's Law of Geographical Information (Taylor and Overton 1991) posits that the need for information tends to be greatest precisely where least is available. Although the main thrust of this criticism is that the most interesting questions in geography invariably lie beyond the realms of digital analysis, it also has implications that are closer to geocomputation practice. Platitudes about data-rich environments and information economies cannot disguise the fact that the sources and operation of many biases in data collection and availability are very poorly understood. A good example is provided by the supplementation and partial replacement of (census-based) geodemographic data with lifestyles datasets. Lifestyles data are richer, more detailed and up-to-date and, arguably, predictively successful, than their geodemographics forebears: yet at the same time they are fundamentally more biased and unscientific. Another example is that discussed by Curran et al (this volume) for whom geocomputation is principally about the processing, classification and management of large datasets: such interpretation is avowedly subjective, and (with regard to urban image classifications in particular) dependent upon the sampling of 'ground truth' observations and *a priori* under-standing of the configuration and intensity of land use. A third and more general point is that in our emergent networked digital world, data sharing is increasingly set to become the norm, and datasets will likely be handled by many users who were not involved with their creation. Geocomputational practice needs to be fully aware of the spatial and other attributes of such datasets, and Goodchild (this volume) describes the ways in which metadata ('data about data') are being developed in this context.

As with data, so also with software. Software is breaking up into modules which can be assembled over networks at will, and interoperability among and between data products and software systems is becoming increasingly the norm. The ways in which software systems handle areal units and diverse data structures further emphasise the inherent subjectivity of even the most apparently 'objective' models of reality that are abstracted within GIS – be this in the creation of data models of reflected radiation (Curran et al, this volume), behavioural analysis of spatial interactions, or geometric transformations of diffusion processes (Cliff and Haggett, this volume). A welcome development in geocomputation literature is the attempt to make general techniques sensitive to context, as in the geographically weighted regression of Fotheringham (1997) and multilevel modelling approaches (e.g. Jones 1991; Orford 1997). This is most clearly evident in the socioeconomic realm (as students familiar with the innumerable critiques of quantitative geography, founded on the notion of subject–object distinction will be aware) but is certainly not restricted to it (Raper 1998). It is not yet clear to what extent this is leading to the development of what might be described as context-sensitive 'machine intelligence', although Clarke's (this volume) discussion of the ways in which computers may be sensitised to regional dialects for voice input provides one example of an operational learning capacity of a machine. Goodchild (this volume) provides a discussion of how idiographic local 'footprints' might be depicted. Extension of this principle to problems such as zone design (Openshaw 1996; this volume) and spatial autocorrelation would seem plausible, although there have been false dawns in the realm of artificial intelligence before.

Taken together, it is clear that there is much in geocomputation that needs to be made sensitive to context. No stage in scientific reasoning – choice of data, software and hardware – can be considered in isolation, yet what is also clear is that the practice of science cannot remain aloof from the far-reaching improvements in digital data infrastructure and interoperable software tools that are now taking place. Geocomputation has been caricatured as uninformed pattern-seeking empiricism in the absence of clear theoretical guidance, yet it clearly is and must be much more than this. New data and new computation mean new realms for machine 'intelligence', but model structures must be transparent and clearly specified if science is to mean anything more than time- and space-contingent prediction. There is awareness of the need to clarify the aims, technicalities and objectives of new approaches (e.g. see Openshaw and Openshaw 1997), yet there is still some way to go: for example, not all will see virtue in Openshaw's (this volume) statement that his geographical analysis machine (GAM) 'had a number of *attractive* features: in particular, it was automated, [and] *prior knowledge or ignorance was rendered equally irrelevant*' (emphasis added). For some this amounts to the 'higher superstition' (Gross and Levitt 1994) that also characterises profoundly non-scientific views of the world, particularly if the resultant patterns only point the finger at spatial surrogates or indeed nothing of real consequence at all.

This discussion has focused on the internal debates about the scope and content of geocomputation, although as geocomputation develops so it will also come under external scrutiny. The broader context and domain of GIS-based analysis is increasingly researched and understood (Pickles 1998), and the emergent dialogues that have developed will not be rehearsed here. Geocomputation does, however, contribute one major additional consideration to this debate, through its focus upon dynamics. Soja (1989, quoted in Pickles 1995) has suggested that GIS has reintroduced to geography 'notions of space as the dead and the inert' and has argued that this impedes understanding. However, the thrust of much of geocomputation is towards the representation of spatial dynamics, as exemplified by the contributions to this book from data (Curran et al, this volume) through to diffusion modelling (Cliff and Haggett, this volume). Soja's own disquiet about the use of the mosaic metaphor of urban land use (Johnston 1998), for example, is in no small measure met by the transition from mosaic to kaleidoscope, in which space may be transformed and/or represented using different geometries (Batty and Longley 1994).

1.3   About this Book

In the next chapter, Couclelis (this volume) discusses whether geocomputation is, or will become, anything more than a grab-bag of techniques and applications, and whether it has the potential to exert any unifying effect upon science. At this point in time, we can only assume that geocomputation is what its researchers and practitioners do, nothing more and nothing less, but that it aspires to much more than unfettered inductivism. This may seem a less than concrete agenda, but, as the contributions to this book show, there are other important themes that characterise the geocomputational approach – such as improvements in the power of computer processing, the volume of data that may be processed, diversity of new data capture devices, and proliferation of new digital data sources.

We began by suggesting that the broad agenda of geocomputation entails devising models, or selective abstractions, of geographical phenomena. Such abstractions may be used to summarise spatial data in numerical form, or to visualise spatial phenomena – and such data models might then be used as inputs to higher level models of system behaviour and dynamics. Abstraction proceeds through devising models of models – each characterised by their different data origins, their diverse (but interoperable) software, their particular field, desk and network-based configurations of hardware – within a powerful, distributed data and software environment.

Following Couclelis' discussion of the foundations to geocomputation, the remainder of this book is structured into four further parts: these address data and their documentation; data exploration and pattern detection; visualisation of virtual and real environments; and space–time dynamics. Lest this structure be taken to imply a rigid chronology of stages in research design, the recursive, iterative and sharing spirit of geocomputation should first be emphasised here, and to this end we will conclude by drawing the often intricate linkage between the recurring themes in the different contributions. If this book were produced using electronic media, extensive hyperlinking might have been used to highlight these links, although we have used extensive cross-referencing throughout the chapters as an analogue copy substitute.

1.3.1   Hardware and Software Cycles

Irrespective of whether technology is viewed as catalyst to or cause of the innovation of geocomputation, what is abundantly clear is that developments in technology are bringing the environment of computation closer to the real world. Clarke (this volume) describes how portable, even wearable computers permit not only vastly enhanced field data capture possibilities, but also greatly open up prospects for real time linkage and decision support – a theme also taken up by Batty et al (this volume: but see Swann 1998 for a discussion of some limitations!). We have argued above that description often precedes theory in science, and Clarke's (this volume) view is unambiguously that the richness of description is in turn directly related to technology – and that the media of truly portable computing are now driving the development of geocomputation. He suggests that the relative importance of hardware over the last 30 years has been cyclical: i.e. in the early stages of GIS the operating systems of computers and limited capabilities of peripheral hardware devices dominated the science of GIS, while the subsequent innovation of low-cost PCs, workstations and desk-top peripherals made research much less dependent upon platform and device. Today, miniaturisation of computer hardware is making ubiquitous, portable computing possible, and constant technological development is leading to an array of hardware innovations – such as the development of hyper-interactive user interfaces – but that this time hardware is infinitely adaptable not just to the software environment but to the real environment in which it to be used.

Over this period, software development has followed a different trajectory. In the early days of GIS, systems were designed for particular hardware configurations, and the move towards device-independent graphics has been a comparatively recent innovation. The broad trend towards application-specific software which can be used across different domains is termed 'interoperability' (Sondheim et al 1998). Allied to this, the innovation of the Internet has brought at the same time both the fragmentation and convergence of software: fragmentation, in that more and more interoperable GIS software can be assembled together across the Net; and convergence in the sense that software is increasingly 'bundled' with hardware and sometimes data into 'desk-top' GIS. With regards to research applications, the trend towards software break-up in some senses brings today's geocomputation researchers full circle – back into the situation of the early quantitative geographers who wrote their own computer code to solve specific research problems.

Taken together, the prospect of ubiquitous hardware, interoperable software and networking in new and novel ways constitutes a panacea for geocomputation, as described by Batty et al (this volume) and Clarke (this volume).

1.3.2 Data

A second recurrent theme through this book concerns the creation, extraction and accessing of useful information from large and frequently complex datasets. It is ironic that as the world becomes more and more data rich, so we are becoming less and less certain about the ways in which we construe meaning from data. And neither is this a characteristic of the socioeconomic realm alone, in which the relativistic honing of meaning to context has become a central tenet of much of GIS-based analysis (Batty and Longley 1996). For Curran et al (this volume), geocomputation is fundamentally relevant to remote sensing because it can be used to manage large datasets (see also Openshaw, this volume) and, through both inductive and deductive reasoning, provides an environment for creating the best data depictions for specified applications. 'Best' is here not defined exclusively in conventional statistical senses – as illustrated by the failure of globally accurate but implausible (at fine scales and in visual terms) Kriged surfaces to represent terrain roughness, for example. Curran et al's contribution is also resonant of the arguments of Cliff and Haggett (this volume), who see geocomputation as central to the geometrical transformation of space – although Curran et al are principally concerned with geometric rectification rather than recasting space into different (application-specific) travel distance or behavioural metrics. A further common theme linking these two chapters is the emphasis upon geometrical transformation and rectification as a precursor to representing spatial dynamics.

A further data-centric theme in the chapters of this book is that the vastly increased quantities of data do not make them any less imperfect, incomplete and error-prone. As such, data models remain beset by errors, ambiguities and uncertainty (Fisher 1998). Elsewhere in physical geography, Hutchinson and Gallant (1998) have written about appropriate representation of process and scale in the creation of digital terrain models, while Brooks and Anderson (this volume) identify data quality as a key issue in comparing the efficacy of conceptual versus physically based models. A common thread between such examples is that poor data quality inevitably leads to poor decisions. But what are 'poor' data, and how may they be identified? For Goodchild (this volume) the focus is not just on the supply of vast volumes of data, but also the need for appropriate metadata (data about data) and appropriate spatial search engines to verify qualitative data characteristics. The creation and maintenance of metadata is one of the crucial emergent management issues of GIS, in that metadata enable us to form a view about the ways in which reality has been simplified in particular data models and to understand the (scientific or other) assumptions that have been invoked in so doing. Of course, they also enable us to tackle a range of more obvious issues, such as those of temporal specificity which are important when diverse datasets are concatenated together.

In analytical terms, what is also interesting about metadata is that they allow documentation of what is distinctive about localities – 'footprints' in Goodchild's terminology – and geocomputational approaches such as geographically weighted regression (Fotheringham 1997) and multilevel modelling (Jones 1991) are allowing us to measure and record what is distinctive about localities in spatial as well as aspatial senses.

1.3.3 Visualisation and Transformation

Understanding data quality, representation and structure is a logical precursor to data modelling in two, three and higher dimensions. Geocomputation is changing not only the range and scope of such transformations, but also the ways in which we are able to interact with them. To Batty et al (this volume), the essence of geocomputation lies in interactive representation and modelling of space. For Clarke (this volume), our new-found abilities to carry out such tasks in real time and in mobile computing environments is moving geocomputation directly into the domain of human experience and reasoning. For most of the other contributors to this book, the environment of geocomputation is important in making possible generalisation from two to three dimensions (e.g. Brooks and Anderson, this volume; Curran et al, this volume), remote (Internet) browsing and interaction as avatars (Batty et al, this volume; Clarke, this volume), visualising local indices of association (Anselin, this volume; Openshaw, this volume), and measuring properties of spatial dependence and contiguity (Anselin, this volume; Curran et al, this volume).

1.3.4 Induction, Deduction and Generalisation

Much is made in Couclelis' discussion (this volume) of the perception of some areas of geocomputation as being fundamentally data led. This view arises in large part through the work of Openshaw, reviewed by him in this book (Openshaw, this volume) in a review which traces the development of computational data mining techniques in geography. A central assumption of much of this work is that machine 'intelligence' can be of greater import than *a priori* reasoning, by virtue of the brute force of permutation and combination – 'might makes right' in this view of geocomputational analysis, and Openshaw presents extensive evidence that such pattern-seeking is more predictively successful than other approaches. Others will feel uneasy with an approach which appears simultaneously to indict human reasoning, scientific practice and procedures of inference as conventionally understood. Moreover, and in the context of Couclelis' view that geocomputation is in need of applied exemplars, this approach appears to date not to have generated substantive findings which have 'stuck': within the realm of epidemiology, medical geographers have been successful in tracing and modelling the geography of disease transmission ever since John Snow (Cliff and Haggett, this volume), yet the substantive findings from geographical explanation machines (GEMs) appear to have become bogged down in questions of equifinality, multiple testing and quality of inference. This might be taken to suggest that such techniques are most appropriately considered to be important prior to the hypothesis-generation stage of research design, or that they might best be fed into deductive reasoning. The latter notion is pursued by Curran et al (this volume) with regard to image classification.

Elsewhere in this book the emphasis is upon the ways in which geocomputation is contributing to geographical science as conventionally practised. Anselin (this volume) discusses a range of developments in the measurement of contiguity/spatial autocorrelation and assignment of spatial weights. More generally, Anselin has developed and applied techniques of exploratory spatial data analysis (ESDA) to visualise spatial distributions in order to identify atypical locations (spatial outliers), to discover spatial clusters, and to suggest different spatial regimes and other forms of spatial heterogeneity.

Such data exploration is in the same spirit as a whole raft of techniques for investigating scenarios, conducting sensitivity analyses, and undertaking probabilistic interpretations, such as those described by Batty et al (this volume), Brooks and Anderson (this volume) and Burrough (this volume). Elsewhere Clarke (this volume) also describes how the availability of raw computing power is enriching established techniques (such as spatial interaction modelling and Monte Carlo simulation) that previously had no analytical or heuristic solution.

1.3.5 Representing and Modelling Spatial Interactions

Throughout the chapters in this book there are numerous examples of the ways in which geocomputation is being used to build dynamic and detailed models of spatial distributions – be these models of spatial distributions as ends in their own right (Curran et al) or as a medium for linking time series in models of dynamic processes (Anselin, this volume; Brooks and Anderson, this volume; Burrough, this volume; Cliff and Haggett, this volume). The linkage of different data models and linkage across time are central to this quest. Most of the applications reviewed in this book are developed in an interactive computing environment, allowing incorporation of feedback from earlier stages in the model-building process.

Good dynamic models require 'a deep understanding of the processes involved, excellent data in large amounts, and sufficient computational power' (Burrough, this volume) and the momentum in dynamic modelling is coming from the assembly of powerful distributed software modules (across networks and based on loose or close coupling) pertaining to both small- and large-scale phenomena and processes (Brooks and Anderson, this volume).

1.4   Conclusions

This book is intended as a primer to a new and rapidly developing area. It will be clear from reading the chapters in this book that there is no clear consensus about how geocomputation is likely to develop, and that some of the chapters are written from quite personal standpoints. As such, readers are to some degree left to make up their own minds as to how the field is likely to develop – and given the present rate and pace of innovation and change, they will not have to wait long to see whether their views are vindicated! The chapters that follow are intended to lay the foundation stones for the future – but inevitably in a speculative rather than a definitive way.

References

Batty M, Longley P A 1994 *Fractal cities: a geometry of form and function.* London and San Diego, Academic Press

Batty M, Longley P A 1996 Analytical GIS: the future. In Longley P A, Batty M (eds) *Spatial analysis: modelling in a GIS environment.* Cambridge, GeoInformation International: 345–52

Birkin M 1995 Customer targeting, geodemographics and lifestyle approaches. In Longley P A, Clarke G P (eds) *GIS for business and service planning.* Cambridge, GeoInformation International: 104–49

Dawkins R 1986 *The blind watchmaker.* London, Longman Scientific and Technical

Elshaw Thrall S, Thrall G I 1998 Desktop GIS. In Longley P A, Goodchild M F, Maguire D J, Rhind D W (eds) *Geographical information systems: principles, techniques, management and applications.* New York, John Wiley, 1: 329–43

Fisher P F 1998 Models of uncertainty in spatial data. In Longley P A, Goodchild M F, Maguire D J, Rhind D W (eds) *Geographical information systems: principles, techniques, management and applications.* New York, John Wiley: 1: 191–204

Fotheringham A S 1997 Trends in quantitative methods 1: stressing the local. *Progress in Human Geography* 21: 88–96

Goodchild M F 1992 Geographical information science. *International Journal of Geographical Information Systems* 6: 31–45

Goodchild M F, Longley P A 1998 The future of GIS and spatial analysis. In Longley P A, Goodchild M F, Maguire D J, Rhind D W (eds) *Geographical information systems: principles, techniques, management and applications.* New York, John Wiley: 1: 567–80

Gross P R, Levitt N 1994 *Higher superstition: the academic left and its quarrel with science.* Baltimore, Johns Hopkins University Press

Haggett P, Cliff A D, Frey A E 1977 *Locational analysis in human geography*, 2nd edition. London, Edward Arnold

Hutchinson M, Gallant J 1998 Representation of terrain. In Longley P A, Goodchild M F, Maguire D J, Rhind D W (eds) *Geographical information systems: principles, techniques, management and applications.* New York, John Wiley, 1: 105–24

Johnston R J 1998 Geography and GIS. In Longley P A, Goodchild M F, Maguire D J, Rhind D W (eds) *Geographical information systems: principles, techniques, management and applications.* New York, John Wiley, 1: 39–47

Jones K 1991 Specifying and estimating multilevel models for geographical research. *Transactions of the Institute of British Geographers* 16: 148–59

Longley P A, Goodchild M F, Maguire D J, Rhind D W 1998 Introduction. In Longley P A, Goodchild M F, Maguire D J, Rhind D W (eds) *Geographical information systems: principles, techniques, management and applications.* New York, John Wiley, 1: 1–20

Macmillan W (ed.) 1989 *Remodelling geography.* Oxford, Blackwell.

Macmillan W 1998 Computing and the science of geography. *University of Oxford Economic Research Group Working Paper* WPG 98–7

Maguire D J 1998 GIS customisation. In Longley P A, Goodchild M F, Maguire D J, Rhind D W (eds) *Geographical information systems: principles, techniques, management and applications.* New York, John Wiley, 1: 357–67

Mandelbrot B B 1986 Multifractals and fractals. *Physics Today* September: 11–12

Openshaw S 1977 A geographical solution to scale and aggregation problems in region-building, partitioning and spatial modelling. *Transactions of the Institute of British Geographers* 2: 459–72

Openshaw S 1996 Developing GIS relevant zone based spatial analysis methods. In Longley P, Batty M (eds) *Spatial analysis: modelling in a GIS environment.* Cambridge, GeoInformation International: 55–73

Openshaw S, Openshaw C 1997 *Artificial intelligence in geography.* Chichester, John Wiley

Openshaw S, Taylor P 1979 A million or so correlation coefficients: three experiments on the modifiable areal unit problem. In Bennett R J, Thrift N J, Wrigley N (eds) *Statistical applications in the spatial sciences*. London, Pion: 127–44

Orford S 1997 *Valuing the built environment: a GIS approach to the hedonic modelling of housing markets*. Unpublished PhD dissertation, University of Bristol

Pickles J 1995 Representation in an electronic age: geography, GIS and democracy. In Pickles J (ed.) *Ground truth: the social implications of geographic information systems*. New York, Guilford Press

Pickles J 1998 Arguments, debates, and dialogues: the GIS–social theory debate and the concern for alternatives. In Longley P A, Goodchild M F, Maguire D J, Rhind D W (eds) *Geographical information systems: principles, techniques, management and applications*. New York, John Wiley, 1: 49–60

Raper J 1998 Spatial representation: the scientist's perspective. In Longley P A, Goodchild M F, Maguire D J, Rhind D W (eds) *Geographical information systems: principles, techniques, management and applications*. New York, John Wiley, 1: 61–70

Shiffer M 1998 Managing public discourse: towards the augmentation of GIS with multimedia. In Longley P A, Goodchild M F, Maguire D J, Rhind D W (eds) *Geographical information systems: principles, techniques, management and applications*. New York, John Wiley, 2: 725–33

Smith N, Rhind D W 1998 Characteristics and sources of framework data. In Longley P A, Goodchild M F, Maguire D J, Rhind D W (eds) *Geographical information systems: principles, techniques, management and applications*. New York, John Wiley, 2: 655–66

Soja E 1989 *Postmodern geographies: the reassertion of space in critical social theory*. London, Verso

Sondheim M, Gardels M, Buehler K 1998 GIS interoperability. In Longley P A, Goodchild M F, Maguire D J, Rhind D W (eds) *Geographical information systems: principles, techniques, management and applications*. New York, John Wiley, 1: 345–56

Swann D 1998 Military applications of GIS. In Longley P A, Goodchild M F, Maguire D J, Rhind D W (eds) *Geographical information systems: principles, techniques, management and applications*. New York, John Wiley, 2: 893–903

Taylor P J, Overton M 1991 Further thoughts on geography and GIS. *Environment and Planning A* 23: 1087–90

Thrift N J 1981 Behavioural geography. In Wrigley N, Bennett R J (eds) *Quantitative geography: a British view*. London, Routledge & Kegan Paul: 352–65

2

Geocomputation in Context

Helen Couclelis

Summary

This paper considers geocomputation in its disciplinary, epistemological and societal contexts. It addresses three groups of questions: (a) the place of geocomputation within geography, in particular in connection with GIS; (b) the epistemological legitimacy of geocomputation as an approach to geographical research and problem solving; (c) geocomputation's societal antecedents and implications. The paper examines the widely held view of geocomputation as a collection of computational techniques and finds it unsatisfactory for several reasons, central among which is the neglect of any systematic connection with spatial theory and the theory of computation. The essay concludes with some of the challenges facing geocomputation in this broader context of disciplinary, theoretical and societal issues.

2.1 Introduction

The provisional working definition for geocomputation adopted in this essay is the eclectic application of computational methods and techniques 'to portray spatial properties, to explain geographical phenomena, and to solve geographical problems'. This paraphrases Dobson's (1983) early attempt to define 'Automated Geography', a concept we would now readily identify with geocomputation as formulated by the organisers of the first three conferences of that name. According to the prospectus of *Geocomputation '97*, the new term and conference series emerged in response to a growing need to bring together the diverse research efforts seeking to capitalise on the vastly expanded 'opportunities for taking a computational approach to the solution of complex geographical problems'. Geocomputation is understood to encompass a wide array of computer-based models and

---

Geocomputation: A Primer. Edited by Paul A Longley, Sue M Brooks, Rachael McDonnell and Bill Macmillan.
© 1998 John Wiley & Sons Ltd.

techniques, many of them derived from the field of artificial intelligence (AI) and the more recently defined area of computational intelligence (CI). These include expert systems, cellular automata, neural networks, fuzzy sets, genetic algorithms, fractal modelling, visualisation and multimedia, exploratory data analysis and data mining, and so on. The announcement of *Geocomputation '98* on the World Wide Web goes further, stating that 'spatial computation represents the convergence of the disciplines of computer science, geography, geomatics, information science, mathematics, and statistics'. There is a suggestion here of something called 'spatial computation' that may go beyond the mere sharing of computational techniques, since no 'convergence of disciplines' has ever been achieved through the use of spreadsheets, statistical software, or graphics packages.

In considering geocomputation in its broader context, the key question this essay addresses is whether geocomputation is to be understood as a new perspective or paradigm in geography and related disciplines, or as a grab-bag of useful computer-based tools. I will assume that the former is at least potentially the case, as there is little scope in talking about the broader context of grab-bags or toolboxes. I have no quarrel with those to whom geocomputation just means the universe of computational techniques applicable to spatial problems: this is probably quite an accurate characterisation of the current state of the art. The question is whether or not we are witnessing the rise of a distinct intellectual approach to the study of geographical space through computation, that is, whether the geocomputation whole will ever be more than the sum of the computational parts.

This paper considers geocomputation in its disciplinary, epistemological and societal contexts. Accordingly, it addresses three groups of questions: (a) the place of geocomputation within geography, in particular in connection with geographical information systems (GIS); (b) the epistemological legitimacy of geocomputation as a scientific approach to geographical research and problem-solving; (c) geocomputation's societal, historical and institutional antecedents and implications. Many of these issues parallel those surrounding computation in general and more specifically, AI. Without neglecting some of the major points common to all aspects of the computational revolution, this essay will focus on the questions more specific to the spatial sciences and geography in particular. The essay concludes with some of the challenges facing geocomputation in this broader context of disciplinary, philosophical and societal issues.

2.2 Geocomputation and Geography

A first group of issues concerns the place of geocomputation within a geographical research and applications environment that is by now widely computerised and data-orientated. Even disregarding pervasive but clearly non-geographical applications (word processing still being by far the most common use of the computers on most geographers' desks), there are very few areas within geography that have no use for – say – a computer-generated map, graphic, or table. More significantly, there are fewer and fewer areas (and not just within geography) that have not been touched by GIS. In some sense we have been 'doing geocomputation' for many years without knowing it. Less trivially, however, it is fair to say that geocomputation has thus far found only limited acceptance within the discipline. More surprisingly, it even has an uneasy relation with mainstream quantitative geography, as evidenced by the relative dearth of geocomputation-orientated articles and topics in the main quantitative geography journals and texts. These statements must be qualified by the ambiguity of what really counts as geocomputation work, this being a fuzzy set rather than a Boolean category. Still, in my mind, there are clear instances of the set as well as marginal ones, and it is the more (proto)typical ones that tend to find the least recognition. My remarks in this essay concern primarily these more representative cases. There is not much point in trying to specify what is 'core geocomputation' and what is not, but let us agree that the core does not include most traditional computer-supported spatial modelling and analysis of spatial data. In particular, it does not include routine GIS research and applications. Indeed, GIS is in many ways the antagonistic big brother who is robbing geocomputation of the recognition it deserves.

Contrasting the development of geocomputation with that of GIS is instructive. While part of geocomputation broadly defined, GIS has enjoyed tremendous popularity and commercial success for nearly two decades, whereas geocomputation is still waiting to take off both as a concept and as actual practice. Novelty is not the explanation, since neither the individual techniques that are part of it, nor the notion of geocomputation in itself, are very new, as may be seen from Dobson's (1983) 'Automated Geography' article and the ensuing discussion forum in the August 1983 issue of *The Professional Geographer*. This section examines the reasons for this discrepancy and the very different roles of geocomputation and GIS within the disciplinary matrix.

The phenomenal success and rapid spread of GIS both within geography and well beyond have been amply documented. GIS courses in universities cannot keep up with student demand. The GIS specialty group is by far the largest within the US Association of American Geographers (AAG), and there are few academic job advertisements in geography that do not require at least some GIS skills. There are dozens of GIS texts, and, it seems, a new GIS-orientated journal is being launched every few months. GIS conferences abound, some of them drawing thousands of participants. The GIS industry has been booming for the past 15 years and its products can be found all over the globe in both rich and poor countries. In the late 1980s the US National Science Foundation (NSF) deemed GIS important enough to become the focus of a national research priority, funding the National Center for Geographic Information and Analysis (NCGIA). Shortly thereafter, the European Science Foundation (ESF) followed suit, supporting the five-year GISDATA programme across European countries.

The reasons for the success of GIS are many, among them of course luck and serendipity, but some of these stand out as defining everything geocomputation has not yet been. These include the early beginnings in a large-scale, very visible, and eminently practical application; the major institutional and commercial interest that these applied origins attracted; the compatibility with the highly respected cartographic tradition in geography; the strong visual orientation; and the intuitive appeal of the overlay model of GIS operation, a time-honoured principle that could be explained in a few words (or pictures) to anyone not already familiar with overlay techniques. In many ways GIS was geography's 'horseless carriage', a technological advance that would allow applied geographers and others to do faster, more comfortably, and better what they had always done. That the horseless carriage would eventually change the structure of the disciplinary landscape was not foreseen at the time, and certainly most GIS pioneers did not make any such claims. Contrast this reassuring background with that of geocomputation: no major applied demonstration project; no significant institutional or commercial interest in its further development and diffusion; no obvious affinity with any of the major intellectual themes in geography and related fields; no characteristic visual or other easily accessible mode of presentation; no unifying, simple to grasp underlying principle; instead, rumbles about a revolution *ante portas*, that are neither substantiated by the past record nor by some coherent vision of things to come (Dobson 1983; 1993; Openshaw and Openshaw 1997).

Adding insult to injury, many of the original proponents of the notion of geocomputation ended up identifying it with GIS a decade down the road. In 1993, the tenth anniversary of the 'Automated Geography' discussions, *The Professional Geographer* invited most of the original and some new contributors to comment on how the emerging perspective had fared since it was first debated in the journal in 1983. It is instructive to read these more recent contributions written after the GIS revolution had become part of the mainstream. In most of them the terms Automated Geography and GIS are used interchangeably (Cromley 1993; Goodchild 1993; Pickles 1993; Sheppard 1993). Even Dobson himself capitulates under the assault of the GIS steamroller: in his words, 'today GIS is often used implicitly as a covering term for the collective technologies . . . that I listed as components of Automated Geography.' (Dobson 1993: 434). A perusal of the two earlier volumes of geocomputation conference proceedings confirms that this is indeed a fairly common view. Most supporters of geocomputation will agree, however, that GIS, widely regarded to be 'geography's piece of the computer revolution', is only a part of a much broader project that has yet to be fully spelled out.

In my view a critical part of the definition of geocomputation is the distinction, routinely overlooked even by experts, between *computation* and *computing*. As a first approximation this is the difference between approaching the computer as an integral component of the modelling of complex systems and processes, versus treating it as an equation-solving and data-handling device. While there may be no clear demarcation line between these two views, most practitioners will agree that a difference exists and that it is important. The following section explores further the significance of that distinction. But even granting that computation in geography goes beyond the mere use of computing, it is not immediately evident why all the disparate research and application areas that make use of certain kinds of computational technique(s) deserve to be pulled together under a new fancy designation (Kellerman 1983). The 'quantitative revolution' in geography earned its name precisely because it went well beyond the introduction of mathematical and statistical techniques to provide a truly new angle to the conceptualisation of geographical space, problems and phenomena (see Cliff and Haggett, this volume). Will geocomputation ever be able to claim as much?

2.3   The Epistemology of Geocomputation

Consideration of the reasons for the slow diffusion of geocomputation relative to GIS and other computer-based methods and tools quickly leads to epistemological questions. A presentation of geocomputation by some of its proponents as theory-free and philosophy-free has not only deterred many potential adopters but is also factually incorrect. Indeed, it is precisely the fact that geocomputation is loaded with unfamiliar and unclarified epistemological underpinnings that accounts for much of the scepticism with which the approach has been received. This section begins by contrasting the apparent epistemological poverty of geocomputation with the rich philosophical and theoretical roots of quantitative geography. It then proceeds to explore the epistemological background of computation, this being the natural area to search for the basis of a theory of geo-*computation*. Last, the significance of the '*geo-*' prefix is questioned in a brief discussion of whether 'geocomputation' may mean something more than merely the use of computational techniques in geography and related fields.

While contrasting geocomputation with GIS helps clarify mostly the practical reasons for the relatively slow development of the former, a comparison with quantitative geography sheds light on the theoretical side of the issue. The comparison is apposite because of the strong affinities between the two perspectives: both make heavy use of geographical data, both emphasise modelling, both involve extensive use of computers, and most proponents of geocomputation have intellectual roots in mainstream quantitative geography. As every undergraduate knows, geography's quantitative revolution was heavily influenced by the tenets of positivist philosophies and the physicalist model of science. The quantitative movement soon coalesced around the set of research objectives and practices deriving from the positivist formulation of the scientific method: observation and measurement, experimental design, hypothesis testing, mathematical modelling and theory development, and the standards of objectivity, replicability, and analytic thinking (Harvey 1969). What held this intellectual culture together, however, was not the methods themselves but the underlying belief that the world studied by human geography, no less than that of physical geography, is amenable to the same kinds of approaches to description and explanation, if not prediction, as the world of physical science. Moreover, positivist geographers were able to define the whole discipline on the basis of one of the most fundamental notions in physics – space. Geometry, the formal science of space, provided the theoretical foundation on which quantitative geography was built, and cemented the identity of a substantial part of the discipline as the *spatial science*.

Whatever one may think of the positivist world view as an epistemological position (and goodness knows the critics have not left unchallenged an iota of it), it did provide its followers with a strong sense of identity and purpose as well as non-negligible amounts of professional pride. Mainstream quantitative geography, like it or not, is a complete intellectual programme: it has a philosophy, a set of accepted research practices, techniques that draw their credibility from being derived from those of hard science, and a widely accepted conceptual framework based on the abstract properties of space. It did not accomplish everything it had set out to do, but it was a mighty good try (Macmillan 1989). Contrast that with geocomputation: no philosophy (and proud of it!), no set of approved practices to help define the standards of acceptable work, a haphazard collection of techniques developed for all sorts of different purposes in a variety of areas with sometimes unproven intellectual credentials, and, last but not least, no obvious conceptual framework to define some alternative vision of what geography may be about, or where it might go. While these traits may appeal to our postmodern longings for deconstruction, it is no wonder that most quantitative geographers – who, incidentally, took to comput-*ing* (and GIS) like ducks to water – continue to hold geocomput-*ation* at arm's length.

It does not have to be that way, for geocomputation is a stepchild with quite distinguished hidden parentage. Computation is the name of a theory – the highly formal *theory of computation* (Minsky 1967; Lewis and Papadimitriou 1981). In contrast with the computer revolution, which made computers part of everyday life, the computational revolution was a conceptual one, the effects of which were felt primarily within the domains of mathematics, theoretical computer science, cognitive science, and some areas in modern physics. Much better known both within other sciences and among the educated public were the widely popularised developments in chaos theory and complex systems (Waldrop 1992), though few people are aware of the close links between these ideas and the formal theory of computation.

The theory of computation is the theory of *effective procedures*. According to Minsky, one of the pioneers of the theory of computation, the notion of effective procedure

> already promises to be as important to the practical culture of modern scientific life as were the ideas of geometry, calculus, or atoms. It is a vital intellectual tool for working with or trying to build models of intricate, complicated systems – be they minds or engineering systems. . . . I believe it is equally valuable for clear thinking about biological, psychological, mathematical, *and (especially) philosophical questions.*
> (Minsky 1967: viii, emphasis added)

In a nutshell, an effective procedure is a step-by-step description of a process that is sufficiently well specified to be carried out by a dumb machine that can follow instructions. The notion is closely related to that of algorithm but has some additional theoretical connotations. This innocuous-looking idea provides the theory of computation with some very profound questions that may be tackled with the tools of modern mathematics and logic as well as philosophy. What processes in the world can be described in such precise manner? What is the appropriate language for describing specific processes? Could any one fixed language deal with all describable processes? Could any one type of machine deal with all possible effective procedures? Can there be well-defined processes that cannot be expressed as effective procedures? Is the operation of the mind an effective procedure? (Minsky 1967: 104). And so on. Thus at the core of the theory of computation is an unlikely combination of concepts that include, in addition to effective procedure, the notions of machine, language and process. Each one of these is addressed by one of the three main branches of the theory of computation: *automata theory* deals with the sequence of increasingly powerful abstract machines, of which the Turing machine is by far the most important; *formal grammar theory* deals with the hierarchy of languages that can be 'understood' by the hierarchy of abstract machines; and the *theory of recursive functions* uses number theory to represent the computation performed by a Turing machine as a numerical encoding. These three perspectives on computation are formally equivalent in the sense that any results obtained in one of them can be translated into corresponding results in the others, even though the different angles yield different kinds of insights and one or the other may be more suitable for specific purposes (Hopcroft and Ullman 1979; Lewis and Papadimitriou 1981). Adherents of geocomputation are familiar with at least one direct application from each of the first two branches: cellular automata from automata theory, and shape grammars from formal grammar theory. I am not aware of any equally obvious example from the theory of recursive functions, although the programming language LISP that was developed specifically for the needs of AI is based on that theory.

There are two potential contributions of the theory of computation to geo-computation as a theoretical project, which I shall call syntactic and semantic. Here I summarise the more detailed argument developed in Couclelis (1986) regarding the theoretical credentials of AI in particular. The syntactic contribution stems from the hierarchy of formal structures illustrated in Figure 2.1. Set theory and logic begot modern algebra. Aspects of modern algebra are specialised in the mathematics of discrete structures, which include graphs, lattices, Boolean algebras, and combinatorics. At the next level down, there are general theories of modelling, such as Zeigler's (1976), which systematise the formal principles of modelling and simulation in any field. These provide a framework for an integrated representation of the three branches of the theory of computation one level down. Below that we have the more familiar specialisations of the theory of computation such as AI, computer architectures, parallel distributed computing, and the different program-ming languages. There are unbroken chains of structure-preserving relations or structural equivalencies connecting the theory of computation with set theory and logic at one end, and with concrete computer programs at the other. Thus, in principle, any valid computational method or algorithm corresponds to a sequence of valid formalisms at higher levels. While the transformation from model to algorithm is generally well understood, the reverse route is only rarely possible. Still, one level up from the computational technique, automata theory or formal grammar theory can help clarify its properties, perhaps turn a messy 'black box' into concise sets of interpretable statements, and detect logical inconsistencies and other problems (Dijksta 1976). Work along these lines would contribute welcome analytic rigour to some of these techniques, reveal unsuspected affinities among disparate problem-solving approaches, and may help bring about a rapprochement with mainstream quantitative geography with potential major benefits for both sides (see Figure 2.1).

The semantic or conceptual contributions of the theory of computation to a theory of geocomputation would stem from three main sources: the ability to manipulate symbols representing arbitrary entities (e.g. shapes or 'objects') rather

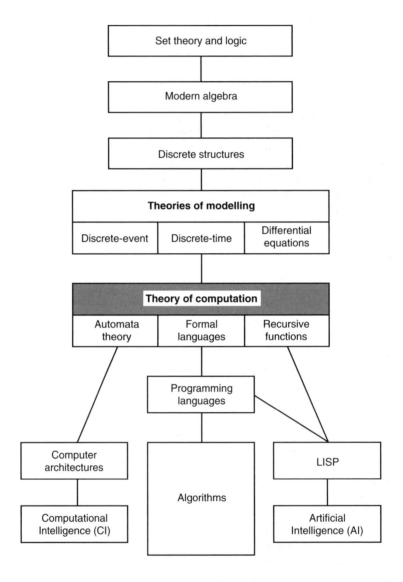

Figure 2.1   *The discrete-structure hierarchy of formalisms (adapted from Couclelis 1986: 4)*

than just numbers; the possibility, reflected in the assignment statement of programming languages, to express qualitative change from one category of things to some other category of things; and the possibility to explore phenomena so complex that there is no better model of them than their full step-by-step description (Couclelis 1986). The logical manipulation of symbols corresponding to arbitrary concepts, the representation of qualitative change, and the simulation of highly complex systems are three areas where the superiority of computational over traditional mathematical and statistical techniques is indisputable. Again here there may be much to be gained from systematic work that would establish these comparative advantages within the context of geocomputation through both theoretical development and practical demonstration. By expanding the realm of modellable phenomena beyond the strictly quantitative there may even be a chance to connect with those who find numbers and measurement in the social domain too limiting a language.

Last but not by no means least in a future theory of geocomputation would be the question of the connection of computation with *space*. The theory of computation is the theory of abstract machines that transform information instead of energy. Adopting computation as a paradigm implies that the underlying notion of machine is appropriate as a fundamental metaphor for the domain in question. Thus AI was built on the premise that the mind operates like an information-processing machine, so that cognitive processes can be modelled as computational processes. More recently the field of artificial life was established on a similar assumption that the processes characterising life, once abstracted from their carbon-and-water realisations, are also analogous to the operation of abstract information-processing machines (Levy 1992). The theoretical challenge for geocomputation is to formulate a model based on the computational notion of machine that justifies the 'geo' prefix. That this may be the case is not at all obvious. The key characteristic of the computational machine model is that it specifies the time-dependent notion of *process*, whereas the theory of scientific geography (and most of the other 'geo'-sciences) is predicated upon the static spatial framework of geometry. In a companion paper I speculate a little further on what the theoretical connection between geography and abstract machines might be, and distinguish two different varieties of geocomputation, 'hard' and 'soft', in analogy with hard and soft AI. I then suggest that hard geocomputation is the computational theory of spatial processes (Couclelis 1998). This sounds like a nice idea, but we are not there yet.

2.4 Geocomputation and Society

Societal and institutional issues raised by geocomputation include its place within a postmodern, globalised information society as well as its status within academia. Just as GIS has had major impacts well beyond geography, geocomputation has the potential to influence a number of other spatial sciences, disciplines and application areas with a spatial component, but also to be noticed beyond the walls of universities and research centres. This potential is based on the fact that geocomputation blends well with several major trends in contemporary society. It is obviously in tune with the computer revolution, and capitalises on the continuing dramatic expansion in computing power and the ubiquity of user-friendly, versatile machines. It has a better chance than stodgy quantitative geography to attract the interest of the coming generation of researchers, who grew up running (and sometimes programming) software as complex as anything we build. It certainly agrees with the strong applications and data orientation of our age, and can help people make better use of information now widely available on the World Wide Web and numerous other easily accessible sources. It fits nicely into the growing push for multidisciplinary work, borrowing techniques from areas far removed from those to which they are normally applied. Busy researchers appreciate the ability to pick and choose just what they need out of the geocomputation toolbox, and most of us do not seem to miss the steep learning curve associated with more systematic approaches. Eclecticist, iconoclastic, computer-minded, and theoretically agnostic, geocomputation is also very much a postmodern intellectual product.

To realise its potential for greater recognition and influence geocomputation must get out of the conference theatre and onto the streets. Here again the histories of the three cognate fields discussed in this essay – GIS, mainstream quantitative geography, and computation yield useful insights. The credibility of these fields within academia went hand in hand with (or in the case of GIS, followed) their practical successes, as – to varying degrees – all three proved to be eminently *useful* in real-world applications. GIS was born to meet concrete institutional needs and continued to grow responding to diverse demands from wide segments of society. At a more modest but still significant scale, land use and transportation planners, public service providers, policy-makers, and numerous interests in the private sector have quantitative geography to thank for spatial interaction and location-allocation models and several other techniques derived from spatial analysis (see, for example, Batty et al, this volume). The theory of computation is at the basis of applied computer science with its myriad of contributions to all aspects of the economy, society and everyday life. More specifically, AI, which started out as a speculation on the function of the mind, has helped improve the performance of a wide array of products of modern technology, from domestic appliances and medical devices to guided missiles. For better or for worse, depending on one's viewpoint, these practical successes have carved a place in society for the corresponding approaches, a place that ensures their continued vitality, both material and intellectual. It is too early to say whether geocomputation will ever have a measurable practical impact of that sort, but, if it does, we can expect an accelerating mutual reinforcement between technical and societal interests.

2.5 Challenges for Geocomputation

Looking ahead there seem to be two kinds of future scenarios for geocomputation, some more likely and evolutionary, others less likely but revolutionary. The former promise a steady modest to moderate expansion of the current trend, whereby geocomputation continues to be viewed as a collection of powerful computational techniques borrowed from a variety of different fields and applicable to diverse geographical problems. There are more researchers experimenting with such techniques, more computational alternatives to traditional methods, more useful algorithms being promoted and improved, more papers appearing in mainstream journals, more geocomputation-orientated sessions at conferences attracting larger academic crowds. There is nothing wrong with that picture except that it is less than what some of us would like to see. It is hard to be content with more of the same when we can glimpse the 'revolutionary' scenarios. However, a Geocomputation

Revolution would need to overcome considerable challenges on several different fronts – practical, theoretical, institutional and societal. Throughout this essay I have hinted at what I believe these to be. In place of a conclusion, here is my personal list of geocomputation's five major challenges.

1. *Geocomputation must develop a few major demonstration projects of obvious practical interest, and persuade commercial companies to invest in and market applications-oriented geocomputation-based products.* The positive feedback loop between the worth of an innovation and the interest of the private sector in it is well known. In principle, the more marketable an idea appears, the more commercial interest there is in polishing, packaging and selling it, which in turn provides more motivation and resources for further basic research and development, which leads to wider demand and more and improved supply. This is the dynamic version of the 'better mousetrap' story. Of course in our less than perfect world many a good mousetrap often goes undiscovered, but people by and large can tell what works for them. Geocomputation *qua* academic pursuit could certainly benefit from a fraction of the commercial spotlight GIS has been enjoying, with all the infusion of talent, research moneys and institutional support that kind of attention brings. A few geocomputation applications have already drawn a fair amount of public attention (Openshaw, this volume; Openshaw et al 1983; 1988), but a couple of isolated projects are not enough to launch the feed–forward loop of commercial development and more research. We need to identify appropriate application niches where geocomputation techniques can be shown to have a competitive edge, and find the entrepreneurial savvy to get the private sector to notice.

2. *Geocomputation must develop scientific standards that will make it fully acceptable to the quantitative side of geography and to other geosciences.* Reliability, transparency, robustness and quality control in geocomputation techniques will greatly enhance the ease of use required to spark commercial and institutional interest in geocomputation applications. At the same time, these are also the qualities that will make such techniques more acceptable to the vast majority of academics working with spatial data who are trained in, and evaluated by, the traditional principles of proper scientific work. Right now quality control is left to the motivation and abilities of individual researchers, many of whom may lack a deeper understanding of the assumptions and limitations of particular techniques. It is necessary to organise, document and clean up the geocomputation toolbox, stick up the warning labels where such are needed, and provide user manuals that people can rely on. This is a non-trivial research agenda considering that the practical logic of computation is very different from that of mathematical and statistical techniques, and is much less likely to have been acquired in quantitative methods classes. This methodological challenge is closely related to the next, theoretical one.

3. *Geocomputation must overcome the current lack of epistemological definition, which makes it appear like little more than a grab-bag of problem-solving techniques of varying degrees of practical utility.* The epistemological agnosticism of geocomputation may seem liberating to some, but if there appears to be no

theory or philosophy it is only because we have not been looking. Approaches and perspectives reflect beliefs about how the world is; techniques and methodologies, about how the world works, and how we may find out about it. The theory and epistemology of computation are both well defined. The former is a demanding mathematical field; the latter reflects a view of the world as complex process and raises profound and original philosophical questions regarding determinism, predictability, the nature of thought, and the limits of the knowable. Not everyone interested in geocomputation can or needs to delve in this domain, but at least some of us should do so from time to time. Applying borrowed techniques professionally is one thing; constructing a new field requires a deeper level of understanding.

4. *Geocomputation must develop a coherent perspective on geographical space – that is, justify the 'geo' prefix.* Computational techniques are used widely today in a broad range of disciplines: how is geo-computation different from econo-, chemo-, or oceano-computation? Or is it more like psycho-computation and bio-computation, which gave rise to highly original and intellectually exciting new fields better known under the name of artificial intelligence and artificial life? In a companion paper I speculate at length on this theme (Couclelis 1998) but the issue must be raised again here. What does geocomputation have to say about geographical space – conversely, what difference does space make to the formal and interpreted properties of computational approaches? There is room here for some full-blooded geocomputational theory, or, more practically, for computational models and languages that are genuinely spatial. We will know that we have established a new approach once we move beyond the piecemeal substitution of computational techniques for more established ones, and begin to formulate truly innovative geographical questions and solutions.

5. *Geocomputation must move in directions that parallel the most intellectually and socially exciting computational developments of the information age.* The context of geocomputation begins and ends with society, and thus the last of the five challenges rejoins the first one. Revolutions, even modest ones, affect not just our ways of doing things but also our ways of thinking. Here again we can learn from the example of GIS. Originally a piece of software for 'doing' land management, it has in recent years provoked both hopes and anxieties relating to a wide array of fundamental societal issues. As an object of reflection and critique, it has attracted the interest of even those geographers who would not normally make use of data-oriented tools or quantitative methods (Pickles 1995). The power of that technology to stimulate thought and debate well beyond the technical is a measure of the success of the GIS revolution (Goodchild 1993). So too geocomputation's growing practical utility may some day find itself engaged in an unanticipated discourse with the societal questions it raises: questions about the wider implications of its forthcoming coupling with the evolving information infrastructure as well as with GIS, questions about the policy uses to which it is put, questions about the kinds of geographies entailed by the computational assumption. We shall then remember fondly the first few geocomputation conferences, and smile at how crisp and technical things had looked back then.

Can geocomputation meet these challenges? Is this all asking too much? . . . It may well be, but we would-be revolutionaries need to keep on dreaming.

References

Couclelis H 1986 Artificial intelligence in geography: conjectures on the shape of things to come. *The Professional Geographer* 38: 1–11

Couclelis H 1998 Geocomputation and space. *Environment and Planning B: Planning and Design* 25 Anniversary Issue (in press)

Cromley R G 1993 Automated geography ten years later. *The Professional Geographer* 45: 442–3

Dijkstra E W 1976 *A discipline of programming*. Englewood Cliffs, Prentice-Hall

Dobson J E 1983 Automated geography. *The Professional Geographer* 35: 135–43

Dobson J E 1993 The geographic revolution: a retrospective on the age of automated geography. *The Professional Geographer* 45: 431–9

Foresman T W 1998 GIS early years and the threads of evolution. In Foresman T W (ed.) *The history of geographic information systems: perspectives from the pioneers.* Upper Saddle River, Prentice-Hall: 3–17

Goodchild M F 1993 Ten years ahead: Dobson's Automated Geography in 1993. *The Professional Geographer* 45: 444–6

Goodchild M F 1998 What next? Reflections from the middle of the growth curve. In Foresman T W (ed.) *The history of geographic information systems: perspectives from the pioneers.* Upper Saddle River, Prentice-Hall: 369–81

Harvey D 1969 *Explanation in geography*. London, Edward Arnold

Hopcroft J E, Ullman J D 1979 *Introduction to automata theory, languages, and computation*. Reading, Addison-Wesley

Kellerman A 1983 Automated geography: what are the real challenges? *The Professional Geographer* 35: 342–3

Macmillan W 1989 *Remodelling geography*. Oxford, Basil Blackwell

Levy S 1992 *Artificial life: the quest for a new creation*. New York, Pantheon Books

Lewis H R, Papadimitriou C H 1981 *Elements of the theory of computation*. Englewood Cliffs, Prentice-Hall

Minsky M L 1967 *Computation: finite and infinite machines*. Englewood Cliffs, Prentice-Hall

Openshaw S, Openshaw C 1997 *Artificial intelligence in geography*. Chichester, John Wiley

Openshaw S, Steadman P, Greene O 1983 *Doomsday: Britain after nuclear attack*. London, Blackwell

Openshaw S, Charlton M, Craft A, Birch J M 1988 An investigation of leukemia clusters by use of a geographical analysis machine. *The Lancet*, Feb. 6: 272–3

Pickles J 1993 Discourse on method and the history of the discipline: reflections on Dobson's (1983) 'Automated Geography'. *The Professional Geographer* 45: 451–5

Pickles J (ed.) 1995 *Ground truth: the social implications of geographic information systems*. New York, Guilford Press

Sheppard E 1993 Automated geography: what kind of geography for what kind of society? *The Professional Geographer* 45: 457–60

Tomlinson R 1998 The Canada Geographic Information System. In Foresman T W (ed.) *The history of geographic information systems: perspectives from the pioneers.* Upper Saddle River, Prentice-Hall: 21–32

Waldrop M M 1992 *Complexity: the emerging science at the edge of order and chaos*. New York, Simon & Schuster

Zeigler B P 1976 *Theory of modelling and simulation*. New York, John Wiley

Part Two
ON DATA AND
DOCUMENTATION

3

Remote Sensing: From Data to Understanding

Paul J Curran, Edward J Milton, Peter M Atkinson and Giles M Foody

Summary

The terrestrial environment interacts differentially with electromagnetic radiation according to its essential physical, chemical and biological properties. This differential interaction is manifest as variability in scattered radiation according to wavelength, location, time, geometries of illumination and observation, and polarisation. If the population of scattered radiation could be measured, then estimation of these essential properties would be straightforward. The only problem would be linking such estimates to the environmental variables of interest. This chapter explores attempts to sample and model the five dimensions of scattered radiation (spectral, spatial, temporal, geometrical, polarisation) and use the results of this sampling to estimate environmental variables of interest. It highlights two issues; first, that relationships between remotely sensed data and the environmental variables of interest are indirect and, second, our ability to estimate these environmental variables is dependent upon our ability to capture a sound representation of the variability in scattered radiation.

3.1   Introduction

Remote sensing is one of the tools we can use to increase our understanding of the environment in space and time (Wickland 1989; Curran et al 1997). To use remotely sensed data for such a task requires them to have the properties of scientific observations, in that they must be internally consistent, relate to other data in a known way and be relevant surrogates for the environment they represent (Mather 1994). Unfortunately, the majority of remotely sensed data do not have

Geocomputation: A Primer. Edited by Paul A Longley, Sue M Brooks, Rachael McDonnell and Bill Macmillan.
© 1998 John Wiley & Sons Ltd.

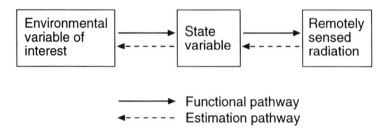

Figure 3.1 *The indirect nature of the relationship between remotely sensed radiation and the environmental variables of interest*

the environmental utility we would hope for and this has served to stimulate a search for such utility that has long pervaded the remote sensing literature. This search has been hampered by two problems that, while *fundamental* (Verstraete et al 1996), are rarely articulated. These two problems are discussed in the text that follows.

Remotely sensed data are nothing more than a measure of radiation after it has interacted with a distant portion of the environment. We usually wish to use this measure of radiation to estimate an environmental variable of interest and it is there that we encounter our first problem, as the link between this radiation measure and the environmental variable of interest is *indirect*. The radiation measure is related directly to those variables that have physically modified it in some way. If the radiation has interacted with, say, a forest then the defining (or 'state') variables would be those physical, chemical and biological characteristics of the soil, vegetation and atmosphere that have changed that radiation. These state variables would include, for example, geometric variables (e.g. tree density, leaf angle distribution) and scattering variables (e.g. internal cellular structure of leaf, phase angle of aerosols). As the 'state variables are the smallest set of variables which are needed to fully describe the physical state of the system under consideration' (Verstraete et al 1996: 203), it follows that *only* state variables can be estimated directly via remote sensing (Figure 3.1). Fortunately, some environmental variables of interest are, in effect, state variables (e.g. leaf area index, LAI) or are closely correlated with state variables (e.g. grass biomass). This relationship between a state variable and the environmental variable of interest is fundamental to our use of remote sensing but is poorly understood.

As state variables define what happened to radiation when it interacted with the environment it follows that *each* state variable can affect a change in the radiation measured. This leads us to our second problem: a single measure of radiation cannot be used to estimate the influence of all the state variables (Verstraete and Pinty 1992). To redress the balance between the number of radiation measurements and the number of state variables (and thereby environmental variables of interest) we need to decrease the effective number of state variables and increase the number of radiation measurements. We can decrease the effective number of state variables by holding the majority constant. There are several means of doing this and the three most common are: (a) restrict analysis to one type of environment (e.g. tropical forest), (b) correct for the effect of the atmosphere and (c) make simplifying assumptions (e.g. spatially constant canopy geometry). We can increase the number of radiation measurements by recording not one, but many such measurements along the spectrum. We can then repeat this at different locations, times, geometries of sensor and illumination, and polarisations (Curran et al 1990; Gerstl 1990). Reduction in the number of state variables under investigation from a large number to a manageable few and an increase in the number of radiation measurements from one to many is fundamental to the use of remote sensing for the estimation of state variables, and hence to the identification of the environmental variables of interest.

From a remote sensing viewpoint we need to, first, be aware of the indirect nature of the link between measured radiation and the environmental variable of interest, second, constrain the number of state variables and, third, increase the number of remotely sensed measurements of radiation. This latter requirement has become a major geocomputational issue in remote sensing as sensors are becoming available that allow us to sample radiation emanating from the environment with ever finer spectral, spatial, temporal and geometric resolutions and with different polarisations. However, the utility of such sampling varies with environment and sensor. To limit the discussion that follows, the environment will be restricted to the terrestrial and the sensors will be restricted to those operating in the optical region (visible to middle infrared) of the electromagnetic spectrum, although many of the issues raised have wider applicability. Formally, sampling the radiative environment can be represented as follows:

$$R = f(\lambda, x, t, \Theta, p) \tag{1}$$

where R refers to radiation; λ indicates that the radiation is spectrally dependent; x refers to the spatial and t the temporal variations; Θ represents the set of angles that specify the geometric configuration of the source of radiation, environmental 'target' and sensor; and p represents the degree to which the radiation is polarised (Curran 1994). The aim of this chapter is to explore the data models that have been developed to use the above equation for the analysis of R recorded in the spectral, spatial and temporal domains and to outline briefly those that are under development for the analysis of R in the geometrical and polarisation domains.

3.2 Utilising $\Delta\lambda$: Data Models in the Spectral Domain

The great spectral variability of terrestrial materials provides the basis for multispectral remote sensing. Data models to classify soils and rocks based on their colour, such as that developed by the Munsell Colour Company (1950), predate quantitative remote sensing by many years. The history of optical remote sensing records a continual quest to recover more of the reflected spectrum, with the expectation that this increase in the number of R measurements as a function of λ will provide ever more information on the state variables and thereby the environmental variables of interest. Initially, the choice of how many spectral bands, and where they were located, was constrained by technical issues of sensor design and telecommunications technology, but by the time the Landsat Thematic Mapper (TM) was designed (Curran 1985), developments in both of these areas allowed greater emphasis to be placed on strong relationships between R and the environmental variables of interest. Thus the Landsat TM has seven spectral bands compared with the four of its predecessor (Landsat Multispectral Scanning System, MSS), and the forthcoming Medium Resolution Imaging Spectrometer (MERIS) sensor has 15 spectral bands, some of which can be 'tuned' precisely to features of interest (Verstraete et al 1998).

The mid-1980s saw the development of imaging spectrometers which essentially record the complete reflected solar spectrum for each pixel in the image (Goetz et al 1985). These have become known as 'hyperspectral' sensors and there are now several airborne hyperspectral sensors operating worldwide (Curran 1994), some as precursors to spaceborne systems. Whilst the spectral richness of such datasets is highly desirable, this comes at a price. First, the volume of data produced per unit area is much higher than that from traditional multispectral sensors, forcing users to choose between discarding some of the spectral bands or reducing the size of area studied. Second, hyperspectral data cannot normally be processed in the same way as multispectral data; even the visualisation of such datasets presents new problems as, instead of a handful of spectral bands, there may be several hundred to choose from. Of these problems, the first is a function of available technology and this is likely to diminish with advances in computing power. The second problem will increase in importance as more and more users gain the computing power to handle hyperspectral data and such data become more common.

Multispectral remotely sensed values of R are usually considered in relation to both λ and x and so can be represented in three different data spaces: spectral space, feature space and image space (Figure 3.2). Of these, spectral space is fundamental since this is what the sensor records from the Earth's surface. Feature space is the essential link between spectral space and image space in that it defines the n-dimensional (D) space within which a particular data model may be created. Image space allows the spatial manifestation of the data model to be investigated in a way which is natural for a human interpreter (see Section 3.3 below).

The dimensionality of feature space usually does not correspond directly to the number of spectral bands sensed. This is because the reflectance of most terrestrial materials changes gradually with wavelength, except for relatively narrow absorption features which are superimposed upon the general trend or continuum spectrum. Multispectral sensors generally have relatively broad bands (50 nm or more) which average out the detail of the R spectrum, whilst a high proportion of the much narrower bands measured by a hyperspectral sensor will be in regions where the R spectrum is relatively featureless. For this reason, it is common to begin analysis by establishing the intrinsic dimensionality of the remotely sensed data by a technique such as principal component analysis (PCA) and visual inspection of the resulting eigenvector images. The standard method of PCA has been modified to make it more applicable to the task of feature extraction in remote sensing. First, a method of 'directed PCA' in which the eigenvectors are aligned optimally with

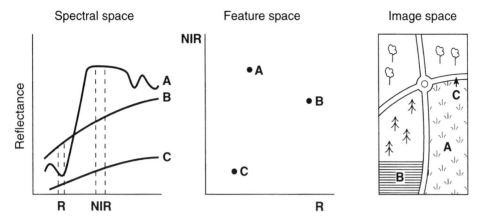

Figure 3.2 *Representation of spectral space, feature space and image space*

general classes of Earth surface materials has been developed (Kauth and Thomas 1976; Jackson 1983). Second, a stage of data pretreatment referred to as 'noise-whitening' has been proposed as a means of ensuring that most information is concentrated in the high-order principal components even when the signal-to-noise ratio of the sensor varies between bands. This PCA procedure is referred to as the 'maximum noise fraction' transform (Green et al 1988), the 'noise adjusted principal components' transform (Lee et al 1990), or (confusingly) as the 'minimum noise fraction' transform (Boardman 1993).

Table 3.1 summarises the intrinsic dimensionality of multispectral R recorded by various remote sensing systems over the last 30 or so years and it is clear from this table that the progression from multispectral sensors with a few spectral bands to hyperspectral sensors with over 200 has not been mirrored by a parallel increase in intrinsic dimensionality. Even the most optimistic assessment of the intrinsic dimensionality of the Airborne Visible/Infrared Imaging Spectrometer (AVIRIS) for example, refers to approximately 15 dimensions (Yuhas et al 1993), whilst many studies report much less than that.

In general, the estimates of intrinsic dimensionality listed in Table 3.1 refer to reconstructions of the continuum spectrum which dominates the signal from soils and vegetation. Very narrow features, such as absorption bands attributable to particular minerals, may be revealed by studying the residuals around the reconstructed continuum spectrum (Gillespie et al 1990) or may be enhanced by limiting the dimensionality analysis to those regions of the spectrum where the absorption features of particular target minerals are known to exist.

Once the intrinsic dimensionality of the data has been established, the cloud of data points from the image can be thought of as defining an n-D shape positioned within the n-D space. The data cloud can be rotated to align more closely with one or more of the bounding axes and it can be projected onto a subset of the dimensions to facilitate viewing in 2-D or 3-D. Care is needed because the 2-D or 3-D representation is but a shadow of the full n-D hypervolume and it is likely that all the intrinsic dimensions contain useful information.

Table 3.1 Estimates of the intrinsic dimensionality of some remotely sensed datasets (refer to text for discussion)

Author(s)	Date	Sensor	Spectral range (µm)	Number of bands used	Intrinsic dimensionality	Description of site/samples and location
Condit	1972	Lab spectrometer	0.32–1.0	100	3–4	Soils: United States
Merembeck et al	1976	MSS	0.5–1.1	4	3	Saudi Arabia
Kauth and Thomas	1976	MSS	0.5–1.1	4	3	Agricultural fields: Illinois
Misra and Wheeler	1977	MSS	0.5–1.1	4	2	Agricultural fields: Kansas
Merembeck and Turner	1980	MSS	0.5–1.1	4	2	NW Pennsylvania
Merembeck and Turner	1980	Skylab S-192	0.41–2.43	12	5	NW Pennsylvania
Crist and Cicone	1984	TM	0.45–2.35	6	3	Vegetation and soils: eastern United States
Smith and Adams	1985	AIS	2.1–2.4	32	6	Hydrothermally altered rocks: Cuprite, Nevada
Huete et al	1985	Multiband radiometer	0.45–2.30	7	3	Cotton canopies over 4 types of soil
Green et al	1988	ATM	0.38–2.35	10	6	Silver Bell district, Arizona
Smith et al	1988	AVIRIS	0.4–2.4	224	5	Owens Valley, California (semi-arid)
Lee et al	1990	GERIS	2.1–2.5	24	8	Western Australia desert
Boardman	1990	GERIS	0.4–2.5	63	3	Hydrothermally altered rocks: Cuprite, Nevada
Smith et al	1990b	TM	0.45–2.35	6	5	Owens Valley, California (semi-arid)
Price	1990	Lab spectrometer	0.55–2.32	178	4	Soils: mostly from the United States

Author	Year	Instrument	Wavelength	Bands	Endmembers	Description
Mackin et al	1990	GERIS	2.0–2.4	21	5–7	Hydrothermally altered rocks: Nevada
Gillespie et al	1990	AVIRIS	0.4–2.4	171	4–5	Owens Valley, California (semi-arid)
Crist	1992	Field spectrometer	0.44–2.33	90	3	Vegetation and soils
Price	1992	Field spectrometer	0.4–2.38	61–206	5–7	Agricultural crops and soils
Roberts et al	1993	AVIRIS	0.4–2.4	224	3	Vegetation and soils: Jasper Ridge, California
Yuhas et al	1993	AVIRIS	0.4–2.4	224	3	Semi-arid grassland (March): NE Colorado
Yuhas et al	1993	AVIRIS	0.4–2.4	224	at least 6	Semi-arid grassland (June): NE Colorado
Yuhas et al	1993	AVIRIS	0.4–2.4	224	5	Semi-arid grassland (August): NE Colorado
Boardman and Kruse	1994	AVIRIS	2.03–2.47	45	3–6	Grapevine Mountains, Nevada
Hurcom et al	1994	AVIRIS	?–1.83	150	3–4	Semi-arid vegetation: Central Spain
Donoghue et al	1994	TM	0.45–2.35	6	4	Intertidal zone: eastern England
Boardman	1994	AVIRIS	?	172	4	Vegetation and soils: Jasper Ridge, California
Harrison et al	1996	Field spectrometer	0.4–2.4	1130	4	Semi-arid vegetation: Central Spain
Milton	1998	CASI	0.41–0.82	10	3	Semi-natural grassland: southern England

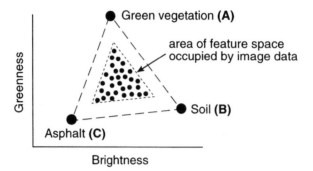

Figure 3.3 *Points A, B and C from Figure 3.2 transformed into 2-D greeness/brightness feature space. The triangular distribution of dots indicates the hypothetical distribution of image data from the scene shown in Figure 3.2, after allowing for mixing between pixels*

If we consider the terrestrial surfaces shown in Figure 3.2, the main dimensions of variation could be captured by a rotation of the feature space such that the new *x*-axis aligned with the axis of bare materials from C (asphalt) to B (bare soil), and the new *y*-axis was orthogonal to that and directed towards the green vegetation point (A). Figure 3.3 shows the location of the same three points in this transformed feature space in which the axes are designated 'greeness' and 'brightness' (Kauth and Thomas 1976). The transformation has reduced the dimensionality of the original spectra and has begun the translation from the measured *R* to the environmental variables of interest.

The data model developed for spectral space, feature space and image space must now be extended to incorporate the imaging process as represented by the sensor model. Most importantly, the relationship between the pure 'endmembers' of the scene, A, B and C on Figures 3.2 and 3.3 must be modified to account for the mixing of objects within pixels, either because of a pixel being located over the boundary between two or more objects or because the instantaneous field-of-view of the sensor is larger than object size. Typically, none of the pixels in the image will be as 'pure' as the areas sampled on the ground, so all the pixels in the scene will be mixed to some extent. This results in distributions of data points such as that shown in Figure 3.3, in which all the image data are within the triangular area defined by the ground samples. The proportion of each of the three classes contained within each pixel is represented by the location of the pixel with respect to the vertices of the circumscribed triangle, which in this case, is approximated by the pure samples measured in the field. This is the basis of the process known as pixel unmixing, which is well suited to hyperspectral data. Most applications of pixel unmixing have assumed that the mixing process is linear, but there is evidence that for some important Earth surface materials the process may be nonlinear and more complex models may be needed (Johnson et al 1983; Roberts et al 1993).

Accurate location of the vertices of the mixing space is important if the estimates of proportions are to be realistic and several methods have been proposed to do this. First, reference endmembers have been chosen to represent the dominant cover types known to be present on the ground (Bierwirth 1990; Harrison et al 1991).

Second, the extremities of scatter plots of the major principal components have been taken to define endmembers and pixels in these regions of the feature space flagged on the image and then visited on the ground to ascribe physical significance to them (Drake and White 1991). Third, Smith et al (1985) describe an approach which combines study of the scatter plots of principal components, constrained by what seems reasonable from field knowledge and spectral matching between image spectra and known samples (e.g. of rocks). This method has since been developed further by Adams and Smith (1986) and Smith et al (1990a). Craig (1990) and Boardman (1993) proposed a method of endmember identification which uses methods of convex geometry to fit a minimum volume shape around the cloud of data points. If the data distribution is sufficiently 'well-behaved' to allow accurate fitting of the convex hull, the vertices of this hull define the spectral endmembers. Tompkins et al (1997) adopted a more explicitly spatial approach in which the mixing equations were modified according to the variance of the data and *a priori* information from the scene.

Pixel unmixing was first explored in the early 1970s (Horwitz et al 1971). However, the intrinsic dimensionality of the remotely sensed data available at that time was insufficient to enable accurate unmixing, since a basic limitation of the method is that the number of endmembers can be no greater than one more than the number of intrinsic dimensions in the data. The resurgence of interest in the technique in the late 1980s was largely because of the increased availability of hyperspectral data with its perceived higher dimensionality. Recent technical advances in sensor performance mean that current state-of-the-art systems, such as AVIRIS, can often produce data with more than six dimensions for many scenes (Yuhas et al 1993). Unfortunately, computational methods to extract information from such data are still poorly developed and those described above for low-dimensional feature spaces become intractable above about 6-D (Boardman 1994). In response to this problem Boardman et al (1995) have developed a technique of partial unmixing, in which all endmembers except the one of interest are collapsed into an unresolved subspace and mixing proceeds in relation to this target : background model. Farrand and Harsanyi (1994) and Harsanyi and Chang (1994) have developed this model further using orthogonal subspace projection (OSP) and the technique of constrained energy minimisation (CEM), while Schaum and Stocker (1994) have discussed the relationship between linear mixing models in general and traditional target detection methods based on statistical criteria, for which the method known as matched filtering is optimal.

Such approaches are well suited to mineral exploration applications in which the aim is to locate exposures of a particular target mineral, but a more general solution is necessary for applications such as land cover mapping. Roberts et al (1998) have developed an approach to this problem using an unmixing technique based on multiple endmembers. In essence, the number and types of endmembers are allowed to vary on a per pixel basis, and each pixel in unmixed in a low-dimensional feature space independent of its neighbours. Using this approach, they showed that over 80 per cent of an AVIRIS image of an area in the Santa Monica Mountains, California, could be unmixed using just two endmembers, although 24 endmembers were mapped across the entire image.

Feature space understanding is fundamental to the development and application of spectral models capable of utilising $\Delta\lambda$ to increase the number of useful measures of R. However, it is not the only data model of importance. The intrinsic dimensionality of a scene is defined in the spectral domain, but it finds its expression in the spatial domain (see Section 3.3), since images are composed of spatially ordered samples of R. For example, image endmembers, which are at the heart of the spectral data model described above, are modified according to the field-of-view of the sensor, as this determines the amount of subpixel mixing which occurs. This is shown in Plate I which plots the mixing spaces and image end-members present in a Compact Airborne Spectrographic Imager (CASI) image from southern England, first using a nominal ground resolution of 2 m and then coarsened to 36 m. In the next section we consider the importance of spatial data models in addressing the explicitly spatial nature of the scene and the information it contains.

3.3   Utilising Δx: Data Models in the Spatial Domain

As we have already seen in Section 3.1 above, a primary objective of remote sensing is to use measures of radiation (state variables) to estimate an environmental variable of interest. The link between this environmental variable of interest and remotely sensed data is often represented by a regression model (for continuous variables) or a statistical classifier (for categorical variables). A first step in moving from data to understanding is the realisation that, in isolation, a remotely sensed pixel value of R is strictly of no value and so contains no information (Atkinson 1995). Information exists only in the relations *between* data, and it is for this reason that we need to transform the remotely sensed data to some standardised measure of R or some environmental variable of interest, whether it be a biophysical variable (e.g. biomass) or a categorical variable (e.g. land cover). Once the data are represented within a framework of standard units to which we can refer, then we can interpret the values in terms of our *a priori* knowledge and extract meaning and thereby new knowledge.

Given the above framework, it is clear that a considerable portion of the useful information in remotely sensed images will reside in the spatial domain within the relations between the pixels in the image. It is curious then that, relative to the research undertaken in the spectral domain (Section 3.2), little attention has been paid to developing explicit tools for exploiting what is a wealth of spatial information. In part, the reason is that the data or information are provided to the user in the form of an image or map, leaving the human brain to observe values of R or some environmental variable of interest and perceive the implicit spatial relations between them. In many ways, this strategy is intuitively sensible because one of the strengths of human perception lies in the ability to process and interpret spatial information through the visual–sensory systems. A secondary reason for the initial sparsity in explicitly spatial research in remote sensing is that while much potential information exists in the spatial domain, there may be considerable redundancy through duplication of relations (Atkinson 1995).

The primarily object-based functioning of the human brain reduces the complexities of the real world to a model world (particularly object-based) in which the redundancy mentioned above is minimised. This object-based approach is increasingly achievable on a computer (Janssen and Molenaar 1995). However, it is not applicable to continuous (e.g. biophysical) environmental variables. In the remainder of this section we explore methods of incorporating spatial information into the remote sensing of both continuous and categorical environmental variables.

3.3.1 Spatial Information for Estimating Continuous Variables

Classical statistical techniques such as regression and maximum likelihood classification act as data transformers converting the pixel values of R in the spectral domain to the domain of the environmental variable of interest and as such take no account of the relations *between* pixels. Such approaches seem wasteful because, unless the image represents white noise, proximate pixels in the image will be correlated. Such correlation is referred to as spatial dependence and this may be interpreted broadly to mean that pixels close together are more likely to be similar than those further apart. Spatial dependence may be represented with one of several structure functions, such as the covariance function, the autocorrelation function, and the variogram.

In this chapter we concentrate on the variogram approach, which is a core geostatistical technique for exploring spatial dependence through explicit reference to a structure function (Journel and Huijbregts 1978; Webster and Oliver 1990). Geostatistics is generally applied within a statistical framework known as the theory of regionalised variables (ReVs) (Matheron 1965; 1971). Essentially, a value of a single pixel is modelled as a realisation of (drawing of a random value from) a random variable (RV) – that is, a histogram of constitutent state variables. A complete image is modelled as a realisation of (drawing multiple random values from) a spatial set of RVs know as a random function (RF) (a spatial set of histograms). The important point is that the RV at one location is dependent on and, therefore, conditioned by neighbouring RVs (Curran and Atkinson 1998).

The statistical correlation between a pair of RFs separated by a given lag, **h**, a vector in both distance and direction, may be expressed with the semivariance, $\gamma(\mathbf{h})$, defined as the average squared difference between the RFs (Curran 1988). The semivariance may be plotted against lag to obtain the variogram, $\gamma(\mathbf{h})$, defined as a parameter of the RF model:

$$\gamma(\mathbf{h}) = 1/2E[\{Z(\mathbf{x}) - Z(\mathbf{x} + \mathbf{h})\}^2] \tag{2}$$

To estimate the variogram from sample data (remotely sensed R or environmental variable of interest) one must compute the semivariance for several discrete lags to obtain the sample variogram. The sample variogram can be estimated for $p(\mathbf{h})$ pairs of observations, $\{z(\mathbf{x}_l), z(\mathbf{x}_l+\mathbf{h}), l = 1, 2, \ldots, p(\mathbf{h})\}$, using:

$$\gamma(\mathbf{h}) = 1/2p(\mathbf{h}) \sum_{l=1}^{p(\mathbf{h})} \{z(\mathbf{x}_l) - z(\mathbf{x}_l + \mathbf{h})\}^2 \tag{3}$$

Note that the variogram is a 1-dimensional function and, therefore, to represent the spatial dependence in different orientations it is necessary to estimate the variogram more than once.

In geostatistics the variogram is used to inform techniques about the exact nature of spatial dependence in the variable of interest. For example, one of the most common techniques is Kriging, a linear technique for optimal unbiased estimation. It is most often applied to interpolate between sample data. In principle, Kriging is straightforward: the estimate is a linear weighted average of proximate sample data. It is unbiased because the weights are forced to sum to one and it is optimal in the sense that the estimation variance s_E^2 or its square root, the standard error, is minimised. Now s_E^2 is minimised by choosing the weights such that they reflect the set of spatial correlations (as represented with the variogram) (a) between the sample data and the estimate and (b) among the sample data themselves. Further, s_E^2 may be estimated for any sample configuration once the variogram is known, allowing sampling to be optimised. Kriging has been applied in remote sensing to optimise field sampling of environmental variables on the ground (Webster et al 1989; Atkinson 1991) and for data reduction (Atkinson et al 1990).

Kriging applies to single variables only and is not, therefore, an alternative to regression and other data transformers. However, it may be extended readily to two or more variables in which case it is referred to as co-Kriging. Co-Kriging allows the estimation of a (primary) variable of interest from a sparse sample of that variable and a more dense sample of other (secondary) correlated variables. It is ideal for remote sensing applications in which the primary (i.e. environmental) variable is often difficult to sample and remotely sensed imagery provides complete cover of the secondary variable. The advantage of co-Kriging over regression is that while, like regression, it accounts for the direct simple correlation between the variables, it also accommodates the spatial dependence in each of the two variables. This is achieved through using the auto-variograms for each variable and also the cross-spatial dependence between them (measured using cross-variograms). Co-Kriging has been used, for example, to map rainfall (Gohin and Langlois 1993), and to design optimal multivariable sampling configurations for field radiometry (Atkinson et al 1992) and airborne multispectral imagery (Atkinson et al 1994).

In recent years, the objective of remote sensing has been extended from spatial estimation to spatial simulation through several geostatistical techniques for stochastic imaging (e.g. sequential Gaussian simulation, indicator simulation and probability field simulation: Deutsch and Journel 1992). The objective of stochastic imaging is to obtain a map which retains the statistical characteristics (in particular the structure function) of the original data, whereas the objective of estimation (including Kriging) is to obtain the best estimate at a single location irrespective of its relations with its neighbours. The importance of stochastic imaging becomes clear when one realises that Kriged and co-Kriged maps of environmental variables could not be replicated by real measurement because the estimated values are smoothed (i.e. they have a reduced variance). Simulated maps could be generated by adding or subtracting stochastic values to the smoothed base in order to create more realistic 'possible realities'. The downside is that the conditionally simulated value at a given location will be only half as accurate as a Kriged (or co-Kriged) estimate. Stochastic imaging has been applied to map continuous variables from remotely sensed imagery and compared directly with regression and co-Kriging (Dungan et al 1994).

3.3.2 Spatial Information for Estimating Categorical Variables

Spatial information has been less readily incorporated into approaches for classification (primarily of land cover) using remotely sensed imagery. The main reason for this would appear to be that variability in land cover does not square easily with the geostatistical concept of a continuous random field. In particular, where the land cover is comprised of homogenous patches of distinct classes (e.g. agricultural fields) then the assumption of a stationary random field (the nature of the variation is the same from place to place) is violated, and thus geostatistical approaches should not be applied.

This problem may be most clearly demonstrated using one of the simplest (and earliest) methods of incorporating spatial information into a classification, namely post-classification smoothing with a low-pass (majority) filter. This filter works well within large homogenous patches (e.g. agricultural fields), but will tend to remove small or linear features of interest (e.g. roads). Similar problems arise when high-pass filters are used to incorporate spatial information such as the local variance or semivariance of R within a moving window. These approaches seek to characterise the nature of the spatial variation around each pixel and to assign a (texture) value to each pixel which might then be incorporated into a multivariate classification (Carr 1996). Yet problems arise with such approaches because of non-stationarity: within large homogenous patches the texture waveband provides useful information, but problems arise at the boundaries between patches and for linear features. More sophisticated approaches which involve classification based on variogram model coefficients (Ramstein and Raffy 1989), or matching the variogram (defined within a moving window) for pixels to be allocated to the variograms for training areas (Miranda and Carr 1994; Carr 1996) are affected by the same problems. Several researchers (e.g. Ryherd and Woodcock 1996) have developed approaches based on the segmentation of *objects* which go some way towards overcoming these stationarity problems.

Van der Meer (1995) has successfully applied traditional geostatistical techniques to classify geological data in which the boundaries are not defined sharply. This has been achieved by using sequential indicator conditional simulation and indicator Kriging to classify geological types from imaging spectrometer data (see Section 3.2). In a non-remote sensing context, Oliver and Webster (1990) have proposed a geostatistical basis for multivariate classification. Their approach involved an unsupervised classifier. The dissimilarities between the values to be classified (a matrix involving all cells) were modified as a function of spatial proximity to each other (using the variogram), with close cells becoming more alike than previously. Similar approaches hold much potential for remote sensing although modification will be necessary to allow supervised classification and application to large datasets. However, the problems of non-stationarity are likely to remain. Other approaches for incorporating spatial information into the classification exist, such as the Gibbs sampler.

In all of the techniques discussed above, whether they be for the estimation or simulation of continuous or categorical environmental variables, the objective is to increase the accuracy of estimation through incorporation of spatial information. For certain applications the extra effort required may not be justified, but for others it will be, particularly where the accuracy of the estimates or simulations are paramount. The quest to utilise the spatial information inherent in remotely sensed imagery to greatest advantage requires and hence provides greater understanding of: (a) the underlying sampling processes (e.g. the effect of spatial resolution on the variogram and, thereby, geostatistical techniques: Atkinson and Curran 1995; 1997); (b) fundamental concepts such as data, information and uncertainty; and (c) the different techniques which may ultimately provide the best solutions.

3.4   Utilising Δt: Data Models in the Temporal Domain

A remotely sensed dataset generally represents a snapshot view of the environment that is rich in spectral and spatial information. A temporal component can be provided by the revisit cycle of the remote sensing (e.g. satellite) system. This component varies from a few hours (e.g. 12 hours for the NOAA Advanced Very High Resolution Radiometer = AVHRR) to several weeks (e.g. 16 days for the Landsat TM). Multitemporal data enable 'holes' in spatial datasets caused, for instance, by cloud cover to be filled (Fuller et al 1994), episodic events to be studied (Ramsey et al 1997), and environmental variables to be monitored (Liu and Massambani 1994; Jensen et al 1997; Lambin 1997). The last-mentioned capability is of particular significance in the context of this chapter, for although it has proved possible to combine $\Delta\lambda$ and Δx as a means of increasing the number of R measurements (Sections 3.2 and 3.3) it has proved difficult to combine $\Delta\lambda$, Δx and Δt in any form of robust model. Therefore, multitemporal analysis will typically involve the use of $\Delta\lambda$, Δx data models for given points in time. Such analysis can increase usefully the measurements of R where the environment under investigation is undergoing predictable (e.g. diurnal, seasonal) changes (Curran et al 1992). However, such an analysis has proved to be of most value when the environment under investigation is undergoing non-predictable changes. In this role multi-temporal remote sensing has proved to be the only practical means of monitoring significant environmental transformations from relatively short-term and localised effects of flooding (Michener and Houhoulis 1997) to globally significant issues such as deforestation (Green and Sussman 1990; Skole and Tucker 1993) and desertification (Tucker et al 1991). Moreover, since the historical archive of satellite sensor data extends to over three decades, the frequency of change that can be detected extends beyond the most transient. Indeed, because of the value of the historical archive, many satellite sensor programmes are designed to ensure data continuity.

The use of multitemporal remotely sensed data is not, however, without its problems. Analysis must account for the effects of temporal variations in a range of factors, such as the viewing geometry (e.g. seasonal variations in solar elevation which may alter the Sun–target–sensor angular geometry and therefore measured spectral response), atmospheric composition and sensor performance, before the apparent observed changes can be confirmed as real. While the quantitative assessment of change can be difficult (Singh 1989), it should at least be possible to derive information on the direction and nature of many changes. Much attention has focused on land cover change, which can be both a cause and consequence of climatic change (Henderson-Sellers 1994; Meyer and Turner 1994). It is also one of the most important global issues, exerting perhaps a more significant effect on the environment than climate change itself (Skole 1996). It is, therefore, important that accurate data on land cover and its dynamics can be derived. There are many methods for mapping and monitoring land cover with remotely sensed data (Singh 1989; Jensen 1996). Often land cover is mapped from remotely sensed data via a conventional supervised image classification and land cover change assessed through sequential map-comparison. Approaches such as these are widely used as attention is focused only on the land cover classes, and many problems attributable to factors such as inter-image differences in atmospheric conditions or calibration are avoided (Singh 1989). The change detection process may be aided with prior knowledge of the conditions or models that describe the transitional processes. For instance, models of vegetation phenology or cropping practices may aid the analysis of multitemporal datasets (Middelkoop and Janssen 1991).

However, many factors influence the accuracy of change detection approaches, most notably the accuracy with which the individual remotely sensed datasets are classified and co-registered (Townshend et al 1992). These in turn are a consequence of many factors including the classes under study (e.g. discrete or continuous classes, degree of spectral separability), the characteristics of the test site (e.g. topography, distribution of possible ground control points) and the techniques used (e.g. classification algorithm, method of geometric transformation). Of particular concern is the accuracy and appropriateness of the individual land cover maps derived from the sequence of remotely sensed data. The accuracy of each classification is a fundamental constraint on its value for the provision of land cover information. Any errors it possesses may also propagate when interrelated with other land cover maps derived at different times. Large errors may therefore be introduced to a multitemporal dataset, even if the individual components are relatively accurate. Often, however, the individual land cover maps contain significant error in their own right. A major source of this error is that in the derivation of a land cover map, whether from field survey or remote sensing, a single class label is allocated to each patch of land. This 'hard' classification process is only appropriate if the classes are discrete and mutually exclusive and thus separable by sharp boundaries (Foody 1996). Often land cover types are continuous. As such the definition of classes, let alone boundaries between them, is contentious. None the less, it is usually possible to identify a general set of characteristics typical of a particular land cover type as defining membership of a class (e.g. the amount of tree cover required to define a forest class) while recognising that cases of this class may vary significantly in their properties and distinctiveness from other classes. Neighbouring ecological communities, for instance, generally inter-grade and are separated by an ecotone which represents a transitional zone that possess characteristics of the various communities involved (Kent et al 1997).

The limit of only one class label per pixel (or other defined mapping unit) imposed by a 'hard' classification provides not only a poor representation of the actual distribution of the land cover but it also constrains the assessment of land cover change to situations which result in a change in class label, implying a complete change in land cover class over the whole area represented by the pixel (Section 3.2). Furthermore, not all important land cover changes involve a conversion of land cover type. Many environmentally significant changes are gradual, and as a result of their subtlety they may go undetected by conventional change detection techniques. Thus even if the accuracy of both the individual land cover classifications and their co-registration were high, a key problem with the use of conventional post-classification comparison methods of change detection is that they will indicate only complete changes in class membership. Thus, for example, a change detection study may reveal that a pixel has been deforested, changing its class label from forest to non-forest. However, only part of the area represented by the pixel may have undergone a change in class membership; the amount of land cover change required to cause a change in class label is a function of many variables but notably the spectral contrast of the classes concerned. Such sub-pixel land cover transformations can be a significant source of error in change detection studies (Skole and Tucker 1993). Additionally, not all land cover changes involve a complete change in class membership, even at a subpixel level, and post-classification change detection methods may be insensitive to these subtle variations in class properties. Changes in land quality and trend, for instance, are often associated with relatively subtle variations such as a change in the proportion of vegetation cover (Sanden et al 1996) which may not be reflected in a full and complete change in class label. These and other partial changes in land cover may also have a significant impact on the environment. For instance, data on the rate of forest degradation would be helpful in reducing uncertainty over the fluxes of carbon between the biosphere and atmosphere (Houghton 1991) and the thicketisation of savannas which may have a major impact on regional photosynthetic activity (Parton et al 1994; Archer et al 1995). With such partial land cover changes the main concern is a change in the properties of the class (e.g. variations in the amount of woody vegetation in a savanna). This land cover change may be considered as a modification rather than a conversion of land cover type (Riebsame et al 1994; Skole 1994) and these need to be detectable remotely.

One approach to the resolution of the problems encountered through the land cover representation derived with a 'hard' classification is to adopt soft or fuzzy mapping approaches and, if necessary, separate classes with fuzzy boundaries, which define variables such as the sharpness and width of the transition area

(Edwards and Lowell 1996; Wang and Hall 1996). Fuzzy classifications may be derived and compared using a range of techniques and used to illustrate land cover types within a pixel (Foody 1996). Although there are relatively few techniques for the analysis and manipulation of fuzzy classifications (Goodchild 1994; Veregin 1996), the simple comparison of fuzzy classifications may enable partial changes to be assessed (Adams et al 1995; Foody and Boyd 1998). As with conventional post-classification change detection methods, the accuracy of the individual classifications is a major determinant on the accuracy of the changes detected. The ability to detect changes, particularly partial changes, is therefore a function of the sensitivity of the measure of the strength of class membership (derived from the fuzzy classification) to the variations in the presence of the classes on the ground. Comparison of fuzzy classifications may, however, enable the detection of relatively small, subtle, land cover changes (Foody and Boyd 1998) (Plate I). Further developments in these and other image processing techniques, which make it easier to accommodate and interpret multitemporal datasets, should further enhance the contribution of remote sensing in studies of environmental change.

3.5 Utilising $\Delta\Theta$ and Δp: Data Models in the Geometrical and Polarisation Domains

For the majority of environments the changes in R as a function of changes in λ, x and t are large and fairly easy to measure. As we try ever harder to estimate environmental variables of interest from remotely sensed data, so the desirability of also using variability in R with $\Delta\Theta$ and Δp increases.

Relatively few studies have used variation in R as a function of viewing geometry or illumination as a means of strengthening the relationship between measurements of R and the environmental variable of interest (Barnsley 1994; Abuelgasm et al 1996; Barnsley et al 1997). In some ways this is surprising as the non-Lambertian (i.e. anisotropic) form of many environmental targets has been both well characterised (Milton et al 1995) and shown to provide a link to key state variables. The strongest links relate to the optical properties of objects (e.g. the reflectance and transmission of leaves, tree trunks, soil) and the spatial distribution of those objects. These state variables are related in turn to a wide range of environmental variables such as surface albedo and roughness (Barnsley et al 1997). The majority of the research to date has concentrated on vegetation canopies (Kimes 1983) that exhibit a large change in R when sensed along the principal plane (the plane joining the Sun, target and sensor). For a closed canopy R is maximised when the Sun is behind the sensor as all leaves are illuminated and no shadow is visible and R is minimised when the Sun is in front of the sensor as a result of self shadowing (Curran 1983). More importantly, this change in R along the principal plane is related to canopy structure at a range of scales from leaf orientation to the distribution of canopy gaps (Blackburn and Milton 1997).

The form of the relationship between R and the geometry of observation and illumination for a whole hemisphere (rather than just the principal plane) has been

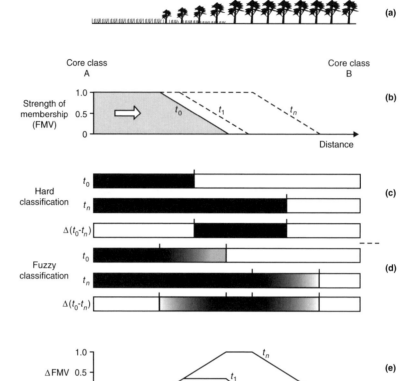

Figure 3.4   *A simplified model of land cover variation along an environmental gradient and the nature of land cover change associated with ecotone displacement. (a) The distribution of two classes separated by a fuzzy boundary along a transect on an environmental gradient, and (b) variation in the strength of class membership (i.e. fuzzy membership value, FMV) associated with one of the classes, class A (grassland), at time t_0. For simplicity it has been assumed that the intergradation of these classes results in a linear decline in membership in the transitional area. Note that outside the transitional area membership is either 1 or 0. A change in environmental conditions over time may result in the migration of the classes and for simplicity it may be assumed that the class migrates uniformly along the transect resulting in the positions indicated by $t_1 \ldots t_n$. (c) The land cover and land cover changes detected by a conventional 'hard' classification. Assuming that each location with a strength of class membership to the grassland class >0.5 is allocated to that class, the distributions at times t_0 and t_n are indicated with a black tone for class A and white for class B. The region of land change between the two times, $\Delta(t_0-t_n)$, is indicated by a black tone. (d) The land cover and land cover changes detected by a fuzzy classification. This shows the extent of class A in black, class B in white and the transitional area as a tonal gradient. The land cover change over the period t_0-t_n is depicted by the grey scale in which black represents a full change in class label and white no change in class label. Note that over the period t_0-t_n the classes and their transitional area migrates onto the land originally forested but that the changes detected may be full or partial. (e) Visualisation of the change in the strength of class membership along the transect associated with the migration. Note at t_1 only partial changes occur but at t_n there is a zone of full land cover change located between regions of partial change. Unlike approaches based on 'hard' classifications, the comparison of fuzzy classifications therefore enables partial and full land cover changes to be observed and characterised (Foody and Boyd 1998)*

used, via geometric models, to infer state variables such as leaf size and inclination (Kimes and Sellers 1985; Li and Strahler 1986; Barnsley et al 1994). For example, multiple view angle (MVA) imagery has been used to provide the multiple measurements of R required to estimate surface albedo (Wanner et al 1997). Our ability to use the geometry of observation and illumination to increase the measurement of R was developed using standard (Milton et al 1995) or in some cases purpose-built field spectroradiometers (Deering and Leone 1986) and the NASA airborne advanced solid-state array spectrometer (ASAS: Irons et al 1991). For example, the ASAS has been used many times to estimate a wide range of state variables (including LAI, canopy cover, gap fraction) in forest canopies (Abuelgasm and Strahler 1994; Johnson 1994; Ranson et al 1994). Although satellite sensors view the Earth at several observation and illumination angles, only the along-track scanning radiometer – 2 (ATSR-2) allows the recording of two look angles (vertical and +55° from zenith) almost simultaneously and at environmentally relevant wavelengths (Harries et al 1995). For vegetated surfaces these images have a high intrinsic dimensionality (Gemmell and Millington 1997) and are capable of estimating a range of state variables that describe canopy structure (Bailey 1997).

The electromagnetic wave recorded at a variety of angles of illumination and observation will be vibrating preferentially in a certain plane. The degree of this preferential vibration is the polarisation (p) of the wave. Sunlight and most reflected radiation is polarised weakly. However, radiation that has been reflected specularly from a smooth surface will be polarised strongly. The degree of polarisation is independent of wavelength and so can, in certain circumstances, be considered as an additive constant to the value of R (Section 3.2 above; Vanderbilt et al 1991). However, the magnitude of this additive constant varies with the characteristics of the surface. For example, the effect of Δp on R can be large for cereal crops with and without shiny flag leaves or soil that is rough and dry as opposed to smooth and wet (Rondeaux and Herman 1991). Since the 1970s, researchers have been trying to measure the proportion of p within measurements of R in space and time, as a means of both estimating state variables such as leaf angle distribution, leaf moisture content, leaf cuticular wax thickness and soil moisture content (Curran 1978; Vanderbilt and Grant 1985) and strengthening the relationship between R and specific state variables (Rondeaux and Vanderbilt 1993). It has proved difficult to link the proportion of R that is polarised to state variables in a consistent manner and so the proportion of R that is polarised has been used along with total R for the estimation of categorical variables such as land cover (Egan 1985; Talmage and Curran 1986).

Of the five variables discussed in this chapter, polarisation has the smallest variability, and as this variability is dependent upon the geometry of the illumination and observation it has not proved practical for many sensors to record it. Most of our understanding of the polarisation properties of terrestrial surfaces has come from placing polarising filters in front of spectroradiometers, cameras and, more recently, matrices of photo-optical detectors (Talmage and Curran 1986). POLDER (POLarization and Directionality of the Earth's Reflectances), a spaceborne imaging polarimeter, has yet to be launched (Deuzé et al 1993), but once in orbit will enable the operational measurement of this elusive fifth variable.

3.6 Summary

We have used this chapter to focus on some of the geocomputational issues that underpin the optical remote sensing of terrestrial surfaces. In doing so we have used an explicit framework that helps to put much contemporary remote sensing research into perspective. We believe that an awareness of this framework is vital for those who hope to move from remotely sensed data to understanding. To promote this framework we will end where we started; not with an equation but with a story (see Box 3.1).

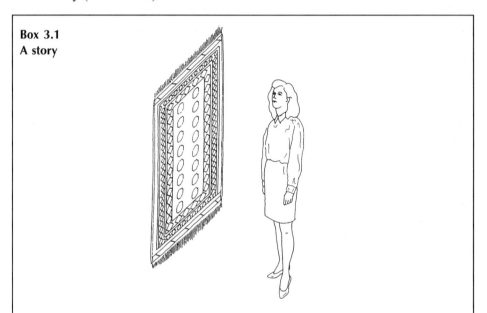

Box 3.1
A story

Once upon a time there was an oriental rug specialist (rs). She was proud of her skills and could value a rug at a glance. The value of an oriental rug is determined by eight 'state of the rug' (state) variables (knot density, material, design, colour range, age, size, condition, country: Bosly 1981) and the rs could see enough of these to derive a value. One day the rs was asked if a single rs valuation (of, say, £500/$800) could be used to estimate the type of environment under which the rug was made; for instance, was it made by nomads? The rs explained the *two* problems she faced. First, while the effects of nomadic production were related indirectly to all of the state variables, the accuracy of such an estimation was dependent primarily on the development of a robust relationship between one state variable (i.e. design) and nomadic production. Second, she would need to minimise the influence of the other seven state variables on the single rs valuation. She realised that she could effectively hold five state variables constant by restricting the scope of her rs valuations by looking at only low knot density, wool-based, single-coloured, small, Middle Eastern rugs. This still left three state variables with only one valuation and so the number of rs variables would need to increase. The rs, therefore, decided to use ultraviolet radiation to estimate age and touched the rug to estimate condition, thus leaving a direct relationship between rs valuation and design and thereby, a usable link between rs valuation and nomadic production.

The rs was then asked if the same rs valuation could be used to estimate the environment after production (i.e. related to the state variable of condition). To this she was able to reply yes, but only if she could also sniff the rug and so add a further rs variable.

References

Abuelgasm A A, Strahler A H 1994 Modelling bidirectional radiance measurements collected by the Advanced Solid-state Array Spectrometer (ASAS) over Oregon Transect conifer forests. *Remote Sensing of Environment* 47: 261–75

Abuelgasm A A, Gopal S, Irons J R, Strahler A H 1996 Classification of ASAS multiangle and multispectral measurements using artificial neural networks. *Remote Sensing of Environment* 57: 79–87

Adams J B, Smith M O 1986 Spectral mixture modelling: A new analysis of rock and soil types at the Viking Lander 1 site. *Journal of Geophysical Research* 9: 8098–112

Adams J B, Sabol D, Kapos V, Filho R A, Roberts D A, Smith M O, Gillespie A R, 1995 Classification of multispectral images based on fractions of endmembers: application to land cover change in the Brazilian Amazon. *Remote Sensing of Environment* 52: 137–54

Archer S, Schimel D S, Holland E A 1995 Mechanisms of shrubland expansion: land use, climate or CO_2. *Climatic Change* 29: 91–9

Atkinson P M 1991 Optimal ground-based sampling for remote sensing investigations: estimating the regional mean. *International Journal of Remote Sensing* 12: 559–67

Atkinson P M 1995 A method for describing quantitatively the information redundancy and error in digital spatial data. In Fisher P (ed.) *Innovations in GIS II*. London, Taylor & Francis: 85–96

Atkinson P M, Curran P J 1995 Defining an optimal size of support for remote sensing investigations. *IEEE Transactions on Geoscience and Remote Sensing* 33: 768–76

Atkinson P M, Curran P J 1997 Choosing an appropriate spatial resolution for remote sensing. *Photogrammetric Engineering and Remote Sensing* 63: 1345–51

Atkinson P M, Curran P J, Webster R 1990 Sampling remotely-sensed imagery for storage retrieval and reconstruction. *Professional Geographer* 42: 345–53

Atkinson P M, Curran P J, Webster R 1992 Co-Kriging with ground-based radiometry. *Remote Sensing of Environment* 41: 45–60

Atkinson P M, Webster R, Curran P J 1994 Co-Kriging with airborne MSS imagery. *Remote Sensing of Environment* 50: 335–45

Bailey P 1997 *Exploring remotely sensed shadow in Amazonian regrowth forests*. Unpublished PhD thesis, University of Southampton

Barnsley M J 1994 Environmental monitoring using multiple-view-angle (MVA) remotely sensed data. In Foody G M, Curran P J (eds) *Environmental remote sensing from regional to global scales*. Chichester, John Wiley: 181–201

Barnsley M J, Strahler A H, Morris K P, Muller J-P 1994 Sampling the surface bidirectional reflectance distribution function (BRDF): 1 Evaluation of current and future satellite sensors. *Remote Sensing Reviews* 8: 271–311

Barnsley M J, Allison D, Lewis P 1997 On the information content of multiple-view-angle images. *International Journal of Remote Sensing* 18: 1937–60

Bierwirth P N 1990 Mineral mapping and vegetation removal via data-calibrated pixel unmixing, using multispectral images. *International Journal of Remote Sensing* 11: 1999–2017

Blackburn G A, Milton E J 1997 An ecological survey of deciduous woodlands using airborne remote sensing and geographical information systems (GIS). *International Journal of Remote Sensing* 18: 1919–35

Boardman J W 1990 Inversion of high spectral resolution data. In Vane G (ed.) *Imaging Spectroscopy of the Terrestrial Environment*, SPIE – the International Society for Optical Engineering 1298: 222–33

Boardman J W 1993 Automating spectral unmixing of AVIRIS data using convex geometry concepts. In Green R O (ed.) *Summaries, 4th Annual JPL Airborne Geoscience Workshop*, Pasadena, California, Jet Propulsion Laboratory: 11–14

Boardman J W 1994 Automating linear mixture analysis of imaging spectrometry data. In

Gomez R B (ed.) *Proceedings, International Symposium on Spectral Sensing Research*, San Diego, US Army Corps of Engineers: 302–9

Boardman J W, Kruse F A 1994 Automated spectral analysis: a geological example using AVIRIS data, North Grapevine Mountains, Nevada. *Proceedings, 10th Thematic Conference on Geologic Remote Sensing*, Ann Arbor, University of Michigan: 501–12

Boardman J W, Kruse F A, Green R O 1995 Mapping target signatures via partial unmixing of AVIRIS data. In Green R O (ed.) *Summaries, 5th Annual JPL Airborne Earth Science Workshop*, Pasadena, California, Jet Propulsion Laboratory: 23–6

Bosly C 1981 *Rugs to riches. An insider's guide to oriental rugs*. London, Allen & Unwin

Carr J R 1996 Spectral and textural classification of single and multiple band digital images. *Computers and Geosciences* 22: 849–65

Condit H R 1972 Application of characteristic vector analysis to the spectral energy distribution of daylight and the spectral reflectance of American soils. *Applied Optics* 11: 74–87

Craig M 1990 Unsupervised unmixing of remotely sensed images. *Proceedings, 5th Australasian Remote Sensing Conference*, Perth, Division of Land Administration: 324–30

Crist E P 1992 Vegetation and soils information in hyperspectral data – an extension of the tasselled cap concept. In Gomez R B (ed.) *Proceedings, International Symposium on Spectral Sensing Research*, Maui, US Army Corps of Engineers: 460–8

Crist E P, Cicone R C 1984 A physically-based transformation of Thematic Mapper data – the TM tasselled cap. *IEEE Transactions on Geoscience and Remote Sensing* 22: 256–63

Curran P J 1978 A photographic method for the recording of polarized visible light for soil surface moisture indications. *Remote Sensing of Environment* 7: 305–22

Curran P J 1983 Multispectral remote sensing for the estimation of green leaf index. *Philosophical Transactions of the Royal Society, London, Series A* 309: 257–70

Curran P J 1985 *Principles of remote sensing*. Harlow, Longman

Curran P J 1988 The semi-variogram in remote sensing: an introduction. *Remote Sensing of Environment* 24: 493–507

Curran P J 1994 Imaging spectrometry. *Progress in Physical Geography* 18: 247–66

Curran P J, Atkinson P M 1998 Geostatistics and remote sensing. *Progress in Physical Geography* 22: 61–78

Curran P J, Foody G M, Kondratyev K Ya, Kozoderov V V, Fedchenko P P 1990 *Remote sensing of soils and vegetation in the USSR*. Taylor & Francis, London

Curran P J, Dungan J L, Gholz H L 1992 Seasonal LAI in slash pine estimated with Landsat TM. *Remote Sensing of Environment* 39: 3–13

Curran P J, Foody G M, van Gardingen P R 1997 Scaling-up. In van Gardingen P R, Foody G M, Curran P J (eds) *Scaling-up: from cell to landscape*. Cambridge, Cambridge University Press: 1–5

Deering D W, Leone P 1986 A sphere scanning radiometer for rapid directional measurements of sky and ground radiance. *Remote Sensing of Environment* 19: 1–24

Deutsch C V, Journel A G 1992 *GSLIB – Geostatistical software library user's guide*. Oxford, Oxford University Press

Deuzé J L, Bréon F M, Deschamps P Y, Devaux C, Podaire A, Roujean J L 1993 Analysis of the POLDER (POLarization and Directionality of Earth's Reflectances) airborne instrument observations over land surfaces. *Remote Sensing of Environment* 45: 137–54

Donoghue D, Reid Thomas D C, Zong Y 1994 Mapping and monitoring the intertidal zone of the east coast of England using remote sensing techniques and coastal monitoring GIS. *MTS Journal* 28: 19–29

Drake N, White K 1991 Linear mixture modelling of Landsat Thematic Mapper data for mapping the distribution and abundance of gypsum in the Tunisian Southern Atlas. *Spatial Data 2000*. Nottingham, Remote Sensing Society: 168–77

Dungan J L, Peterson D L, Curran P J 1994 Alternative approaches for mapping vegetation quantities using ground and image data. In Michener W, Brunt J, Stafford S (eds) *Environmental information management and analysis: ecosystem to global scales*. London, Taylor & Francis: 237–61

Edwards G, Lowell K E 1996. Modelling uncertainty in photointerpreted boundaries. *Photogrammetric Engineering and Remote Sensing* 62: 377–91

Egan W G 1985 *Photogrammetry and polarization in remote sensing*. New York, Elsevier

Farrand W H, Harsanyi, J C 1994 Mapping distributed geological and botanical targets through constrained energy minimization. *Proceedings, 10th Thematic Conference on Geologic Remote Sensing*, Ann Arbor, University of Michigan: 514–24

Foody G M 1996 Approaches for the production and evaluation of fuzzy land cover classifications from remotely sensed data. *International Journal of Remote Sensing* 17: 1317–40

Foody G M, Boyd D S 1998 Detection of partial land cover change associated with the migration of inter-class transitional zones. *International Journal of Remote Sensing* (in press)

Fuller R M, Groom G B, Wallis S M 1994 The availability of Landsat TM images of Great Britain. *International Journal of Remote Sensing* 15: 1357–62

Gemmell F M, Millington A C 1997 Initial assessment of ATSR-2 data structure for land surface applications. *International Journal of Remote Sensing* 18: 461–6

Gerstl S A W 1990 Physics concepts of optical and radar reflectance signatures: a summary review. *International Journal of Remote Sensing* 11: 1109–17

Gillespie A R, Smith M O, Adams J B, Willis S C, Fischer A F, Sabol D E 1990 Interpretation of residual images: Spectral mixture analysis of AVIRIS images, Owens Valley, California. In Green R O (ed.) *Proceedings, 2nd Airborne Visible/Infrared Imaging Spectrometer (AVIRIS) Workshop*, Pasadena, California, Jet Propulsion Laboratory: 243–70

Goetz A F H, Vane G, Solomon J E, Rock B N 1985 Imaging spectrometry for Earth remote sensing. *Science* 228: 1147–53

Gohin F, Langlois G 1993 Using geostatistics to merge *in situ* measurements and remotely-sensed observations of sea surface temperature. *International Journal of Remote Sensing* 14: 9–19

Goodchild M F 1994 Integrating GIS and remote sensing for vegetation analysis and modelling: methodological issues. *Journal of Vegetation Science* 5: 615–26

Green A A, Berman M, Switzer P, Craig M 1988 A transformation for ordering multispectral data in terms of image quality with implications for noise removal. *IEEE Transactions on Geoscience and Remote Sensing* 26: 65–74

Green G M, Sussman R 1990, Deforestation history of the eastern rainforest of Madagascar from satellite images. *Science* 248: 212–15

Harries J E, Llewellyn-Jones D T, Mutlow C T, Murray M J, Barton I J, Prata A J 1995 The ATSR-programme: instruments, data and science. In Mather P M (ed.) *TERRA-2: Understanding the terrestrial environment. Remote sensing data systems and networks.* Chichester, John Wiley: 19–28

Harrison A R, Hurcom S J, Taberner M 1991 Examining spectral mixture modelling using *in situ* spectroradiometric measurements. *Spatial Data 2000*, Nottingham, Remote Sensing Society: 242–9

Harrison A R, Melia J, Bastida J, Gandia S, Gilabert M A, Hurcom S J, Lopez Buendia A, Taberner M and Younis M T 1996 Remote sensing of Mediterranean vegetation and surface lithology. In Brandt J and Thornes J B (eds) *Mediterranean desertification and land use*. Chichester, John Wiley: 493–541

Harsanyi J C, Chang C-I 1994 Hyperspectral image classification and dimensionality reduction: an orthogonal subspace projection approach. *IEEE Transactions on Geoscience and Remote Sensing* 32: 779–85

Henderson-Sellers A 1994 Land-use change and climate. *Land degradation and rehabilitation* 5: 107–26

Horwitz H M, Nalepka R F, Hyde P D, Morgenstern J P 1971 Estimating the proportions of objects within a single resolution element of a multispectral scanner. *Proceedings, 7th International Symposium on Remote Sensing of Environment*, Ann Arbor, University of Michigan: 1307–20

Houghton R A 1991 Releases of carbon to the atmosphere from degradation of forests in tropical Asia. *Canadian Journal of Forest Research* 21: 132–42

Huete A R, Jackson R D, Post D F 1985 Spectral response of a plant canopy with different soil backgrounds. *Remote Sensing of Environment* 17: 37–53

Hurcom S J, Harrison A R, Taberner M 1994 Factor analysis of semi-arid vegetation response using AVIRIS and airborne video data. *Proceedings, 1st International Airborne Remote Sensing Conference and Exhibition*, Ann Arbor, University of Michigan: 424–36

Irons J R, Ranson K J, Williams D L, Irish R R, Huegel F G 1991 An off nadir pointing imaging spectroradiometer for terrestrial ecosystem studies. *IEEE Transactions on Geoscience and Remote Sensing* 29: 66–74

Jackson R D 1983 Spectral indices in n-space. *Remote Sensing of Environment* 13: 409–21

Janssen L L F, Molenaar M 1995 Terrain objects, their dynamics and their monitoring by the integration of GIS and remote sensing. *IEEE Transactions on Geoscience and Remote Sensing* 33: 749–58

Jensen J R 1996 *Introductory digital image processing. A remote sensing perspective*, 2nd edition. New Jersey, Prentice-Hall

Jensen J R, Huang X, Mackey H E 1997 Remote sensing of successional changes in wetland vegetation as monitored during a four-year drawdown of a former cooling lake. *Applied Geographic Studies* 1: 31–44

Johnson L F 1994 Multiple view zenith angle observations of reflectance from Ponderosa pine stands. *International Journal of Remote Sensing* 15: 3859–65

Johnson P E, Smith M O, Taylor-George S, Adams J B 1983 A semiempirical method for analysis of the reflectance spectra of binary mineral mixtures. *Journal of Geophysical Research* 88: 3557–61

Journel A G, Huijbregts C J 1978 *Mining geostatistics*. London, Academic Press

Kauth R J, Thomas G S 1976 The Tasselled Cap – a graphic description of the spectral–temporal development of agricultural crops as seen by Landsat. *Proceedings, 3rd Symposium on Machine Processing of Remotely-sensed Data*, Lafayette, University of Indiana: 41–9

Kent M, Gill W J, Weaver R E, Armitage R P 1997 Landscape and plant community boundaries in biogeography. *Progress in Physical Geography* 21: 315–53

Kimes D S 1983 Dynamics of directional reflectance factor distributions for vegetation canopies. *Applied Optics* 22: 1364–72

Kimes D S, Sellers P J 1985 Inferring hemispherical reflectance of the Earth's surface for global energy budgets from remotely sensed nadir or directional radiance values. *Remote Sensing of Environment* 18: 205–23

Lambin E F 1997 Modelling and monitoring land-cover change processes in tropical regions. *Progress in Physical Geography* 21: 375–93

Lee J A, Woodyatt S, Berman M 1990 Enhancement of high spectral resolution remote sensing data by a noise-adjusted principal components transform. *IEEE Transactions on Geoscience and Remote Sensing* 28: 295–304

Li X, Strahler A H 1986 Geometric-optical bidirectional reflectance modelling of a conifer forest canopy. *IEEE Transactions on Geoscience and Remote Sensing* 26: 276–92

Liu W T H, Massambani O 1994 Satellite recorded vegetation response to drought in Brazil. *International Journal of Climatology* 14: 343–54

Mackin S, Drake N A, Settle J J, Rotfuss H 1990 Towards automated mapping of imaging spectrometry data using an expert system and linear mixture model. *Proceedings, 5th Australasian Remote Sensing Conference*, Perth, Divison of Land Administration: 352–60

Mather P M 1994 Earth observation data – or information? In Foody G M, Curran P J (eds) *Environmental remote sensing from regional to global scales*. Chichester, John Wiley: 202–13

Matheron G 1965 *Les variables régionalisées et leur estimation*. Paris, Masson

Matheron G 1971 *The theory of regionalized variables and its applications*. Fontainebleau, Centre de Morphologie Mathématique de Fontainebleau

Merembeck B F, Turner B J 1980 Directed canonical analysis and the performance of classifiers under its associated linear transformation. *IEEE Transactions on Geoscience and Remote Sensing* 18: 190–6

Merembeck B F, Yates Borden F, Podwysocki M H, Applegate D N 1976 Application of canonical analysis to multispectral scanner data. *Proceedings, 14th Symposium on Application of Computer Methods in the Mineral Industries*, Philadelphia, Pennsylvania State University: 867–79

Meyer W B, Turner II B L (eds) 1994 *Changes in land use and land cover: a global perspective*. Cambridge, Cambridge University Press

Michener W K, Houhoulis P F 1997 Detection of vegetation changes associated with extensive flooding in a forested ecosystem. *Photogrammetric Engineering and Remote Sensing* 63: 1363–74

Middelkoop H, Janssen L L F 1991 Implementation of temporal relationships in knowledge based classification of satellite images. *Photogrammetric Engineering and Remote Sensing* 57: 937–45

Milton E J 1998 Image endmembers and the scene model. *Canadian Journal of Remote Sensing* (submitted)

Milton E J, Rollin E M, Emery D R 1995 Advances in field spectrometry. In Danson F M, Plummer S E (eds) *Advances in environmental remote sensing*. Chichester, John Wiley: 9–32

Miranda F P, Carr J R 1994 Application of the semivariogram textural classifier (STC) for vegetation discrimination using SIR-B data on the Guiana Shield, northwestern Brazil. *Remote Sensing Reviews* 10: 155–68

Misra P N, Wheeler S G 1977 Landsat data from agricultural sites: crop signature analysis. *Proceedings, 11th International Symposium on Remote Sensing of Environment*, Ann Arbor, University of Michigan: 1473–82

Munsell Colour Company 1950 *Munsell soil colour charts*. Baltimore, Kollomorgen Corporation

Oliver M A, Webster R 1990 A geostatistical basis for spatial weighting in multivariate classification. *Mathematical Geology* 21: 15–35

Parton W J, Ojima D S, Schimel D S 1994 Environmental change in grasslands: Assessment using models. *Climatic Change* 28: 111–41

Price J C 1990 On the information content of soil reflectance spectra. *Remote Sensing of Environment* 33: 113–21

Price J C 1992 Variability of high-resolution crop reflectance spectra. *International Journal of Remote Sensing* 13: 2593–610

Ramsey III E W, Chappell D K, Baldwin D G 1997 AVHRR imagery used to identify hurricane damage in a forested wetland of Louisiana. *Photogrammetric Engineering and Remote Sensing* 63: 293–7

Ramstein G, Raffy M 1989 Analysis of the structure of radiometric remotely sensed images. *International Journal of Remote Sensing* 10: 1049–73

Ranson K J, Irons J R, Williams D L 1994 Multiple bidirectional reflectance of northern forest canopies with the Advanced Solid-state Array Spectrometer (ASAS). *Remote Sensing of Environment* 47: 276–89

Riebsame W E, Meyer W B, Turner II B L 1994 Modeling land use and cover as part of global environmental change. *Climatic Change* 28: 45–64

Roberts D A, Smith M O, Adams J B 1993 Green vegetation, non-photosynthetic vegetation and soils in AVIRIS data. *Remote Sensing of Environment* 44: 255–69

Roberts D A, Gardner M, Church R, Ustin S, Scheer G, Green R O 1998 Mapping chaparral in the Santa Monica mountains using multiple endmember spectral mixture models. *Remote Sensing of Environment* (in press)

Rondeaux G, Herman M 1991 Specular and diffuse components of plant canopy reflectance at the wavelengths of 550 and 630 nm. *Proceedings, 5th International Colloquium: Physical Measurements and Signatures in Remote Sensing*. ESA SP319, Noordwijk, European Space Agency: 447–52

Rondeaux G, Vanderbilt V C 1993 Specularly modified vegetation indices to estimate photosynthetic activity. *International Journal of Remote Sensing* 14: 1815–23

Ryherd S, Woodcock C 1996 Combining spectral and texture data in the segmentation of remotely sensed images. *Photogrammetric Engineering and Remote Sensing* 62: 181–94

Sanden E M, Britton C M, Everitt J H 1996 Total ground-cover estimates from corrected scene brightness measurements. *Photogrammetric Engineering and Remote Sensing* 62: 147–50

Schaum A, Stocker A 1994 Subpixel detection methods: spectral unmixing, correlation processing, and when they are appropriate. In Gomez R B (ed.) *Proceedings, International Symposium on Spectral Sensing Research*, San Diego, US Army Corps of Engineers: 278–86

Singh A 1989 Digital change detection techniques using remotely sensed data. *International Journal of Remote Sensing* 10: 989–1003

Skole D L 1994 Data on global land-cover change: acquisition, assessment and analysis. In Meyer W B, Turner B L II (eds) *Changes in land use and land cover: a global perspective*, Cambridge, Cambridge University Press: 437–71

Skole D 1996 Land use and land cover change: an analysis. *IGBP Newsletter* 25: 1–4

Skole D, Tucker C 1993 Tropical deforestation and habitat fragmentation in the Amazon: satellite data from 1978–1988. *Science* 260: 1905–10

Smith M O, Adams J B 1985 Interpretation of AIS images of Cuprite, Nevada using constraints of spectral mixtures. *Proceedings, Airborne Imaging Spectrometer Data Analysis Workshop*, Pasadena, California, Jet Propulsion Laboratory: 62–8

Smith M O, Johnson P E, Adams J B 1985 Quantitative determination of mineral types and abundances from reflectance spectra using principal components analysis. *Journal of Geophysical Research* 90: 797–804

Smith M O, Adams J B, Gillespie A R 1988 *Evaluation and calibration of AVIRIS test-flight data: Owens Valley, CA*. Final Report NASA Contract No. NAGW – 85 Seattle, University of Washington, 1135pp

Smith M O, Adams J B, Gillespie A R 1990a Reference endmembers for spectral mixture analysis. *Proceedings, 5th Australasian Remote Sensing Conference*, Perth, Division of Land Administraton: 331–40

Smith M O, Ustin S L, Adams J B, Gillespie A R 1990b Vegetation in deserts: I. A regional measure of abundance from multispectral images. *Remote Sensing of Environment* 31: 1–26

Talmage D A, Curran P J 1986 Remote sensing using partially polarized light. *International Journal of Remote Sensing* 7: 47–64

Tompkins S, Mustard J F, Pieters C M, Forsyth D W 1997 Optimization of endmembers for spectral mixture analysis. *Remote Sensing of Environment* 59: 472–89

Townshend J R G, Justice C O, Gurney C, McManus J 1992 The impact of misregistration on change detection. *IEEE Transactions on Geoscience and Remote Sensing* 30: 1054–60

Tucker C J, Dregne H E, Newcombe W W 1991 Expansion and contraction of the Sahara Desert from 1980 to 1990. *Science* 253: 299–301

van der Meer F 1995 Estimating and simulating the degree of serpentinization of peridotites using hyperspectral remotely sensed imagery. *Nonrenewable Resources* 4: 84–97

Vanderbilt V C, Grant L 1985 Plant canopy specular reflectance model. *IEEE Transactions on Geoscience and Remote Sensing* 23: 722–30

Vanderbilt V C, Grant L, Ustin S L 1991 Polarization of light by vegetation. In Myneni R B, Ross J (eds) *Photon–vegetation interactions*. Berlin, Springer: 194–228

Veregin H 1996 Error propagation through the buffer operation for probability surfaces. *Photogrammetric Engineering and Remote Sensing* 62: 419–28

Verstraete M M, Pinty B 1992 Extracting surface properties from satellite data in the visible and near-infrared wavelengths. In Mather P M (ed.) *TERRA-2: understanding the terrestrial environment*. London, Taylor & Francis: 203–19

Verstraete M M, Pinty B, Myeni R B 1996 Potential and limitations of information extraction on the terrestrial biosphere from satellite remote sensing. *Remote Sensing of Environment* 58: 201–14

Verstraete M M, Pinty B, Curran P J 1998 MERIS potential for land applications. *International Journal of Remote Sensing* (in press)

Wang F, Hall G B 1996 Fuzzy representation of geographical boundaries in GIS. *International Journal of Geographical Information Systems* 10: 573–90

Wanner W, Strahler A H, Hu B, Lewis P, Muller J-P, Li X, Schaaf C L B, Barnsley M J 1997 Global retrieval of bidirectional reflectance and albedo over land from EOS MODIS and MISR data: theory and algorithm. *Journal of Geophysical Research – Atmospheres* 102: 17143–61

Webster R, Oliver M A 1990 *Statistical methods for soil and land resources survey*. Oxford, Oxford University Press

Webster R, Curran P J, Munden J W 1989 Spatial correlation in reflected radiation from the ground and its implications for sampling and mapping by ground-based radiometry. *Remote Sensing of Environment* 29: 67–78

Wickland D E 1989 Future directions for remote sensing in terrestrial ecological research. In Asrar G (ed.) *Theory and applications of optical remote sensing*. New York, John Wiley: 691–724

Yuhas R H, Boardman J W, Goetz A F H 1993 Determination of semi-arid landscape endmembers and seasonal trends using convex geometry spectral unmixing techniques. In Green R O (ed.) *Summaries, 4th Annual JPL Airborne Geoscience Workshop*, Pasadena, California, Jet Propulsion Laboratory: 205–8

4

Different Data Sources and Diverse Data Structures: Metadata and Other Solutions

Michael F Goodchild

Summary

Geocomputation almost by definition requires access to large quantities of geographical data, and it is increasingly common for such data to be supplied using technologies that support search and retrieval over distributed archives, such as the World Wide Web. It is essential therefore that it be possible to define the characteristics of needed data; to search for suitable sources among archives scattered over a potentially vast distributed network; to evaluate the fitness of a given dataset for use; and to retrieve it successfully. These stages require the development of an array of tools, and associated standards and protocols. The term 'metadata' is commonly used to refer to languages designed for the description of the contents of a dataset, to facilitate its discovery and evaluation by a search engine, as well as its successful transmission and opening by the user's application. In the area of geographical data, the most widely known metadata standards are the Content Standards for Digital Geospatial Metadata, developed and implemented by the US Federal Geographic Data Committee.

4.1 Introduction

Geocomputation has a large appetite for data. While the literature on cellular automata and artificial life shows that it is possible to build interesting simulations within an undifferentiated spatial frame with virtually no input except for model

Geocomputation: A Primer. Edited by Paul A Longley, Sue M Brooks, Rachael McDonnell and Bill Macmillan.
© 1998 John Wiley & Sons Ltd.

parameters, geocomputation focuses on modelling processes on geographical land-scapes that can be sharply differentiated. Its processes respond to assorted boundary and initial conditions, and these must be represented therefore by input of appro-priate geographical data. The parameters of geocomputational models may also be spatially variable, and must be represented with potentially extensive input data. In both of these cases the data serve to differentiate the geographical landscape, and are therefore geographical in the traditional sense, representing the variation of con-ditions or attributes over geographical space: in general, $f(x,y)$, where x and y are positional variables and f is an attribute. Geocomputation is similarly a heavy producer of data, and requires tools for the analysis and display of voluminous simulation results.

The geographical landscape is inherently complex, since x and y define a con-tinuous frame, and its representation can in principle require an infinite amount of information. But the processes being modelled in geocomputation are likely to have inherent *scale*, meaning that there exists some linear measure P that is a property of the process, such that variation over distances less than P has effectively no impact on the outcome of the process, and therefore need not be input (or *resolved*, to use a term common in the modelling community). The literature on physical environ-mental processes contains many attempts to define P both approximately and precisely for specific processes (see, for example, Delcourt et al 1983; Rosswall et al 1988; Ehleringer and Field 1993).

Although the literature of geographical information systems (GIS) has recently begun to include discussion of many types of data, including multimedia (see, for example, Batty et al, this volume; Clarke, this volume; Raper 1997), it seems reasonable to assume that in almost all cases the input to geocomputational models will be strictly geographical, and therefore will follow one of three classes of con-ceptual data models: discrete objects, fields, or their equivalents on 1-dimensional networks embedded in two or three spatial dimensions (Goodchild 1992). This raises one specific issue, however: the set of standard GIS data models does not include all of those data models commonly used in numerical modelling in space. Specifically, while finite-difference modelling (see Burrough, this volume) uses discrete spatial elements that are recognisable as raster data models in GIS, finite-element modelling (FEM) makes use of grids or *meshes* that do not have recog-nisable equivalents in mainstream GIS software. GIS has yet to recognise the importance of representing a field using quadratic functions over triangular elements (GIS TIN implementations use only linear functions), or of polynomial functions over quadrilateral elements (for a review of FEM mesh techniques see, for example, George 1991; Knupp and Steinberg 1993).

Implicit in much discussion of geographical data is the notion of *sharing* (Onsrud and Rushton 1995). Many types of geographical data are collected for very broadly defined purposes, and are widely disseminated and used. A distinction is often made between *framework* data and other types (NRC 1995): framework geographi-cal data are defined as having general use for the purposes of positioning, and for construction of other, more specialised data that can be referenced to them. Many other types of geographical data are also collected for diverse uses, and the processes of collection and use of data can be widely separated geographically, in time, and by discipline – for example, the information contained in a soil database may be collected by a soil scientist, but used by a meteorologist in a model of atmosphere–soil moisture transfer. Remote sensing is a major source of geo-graphical data (Curran et al, this volume), and here also the funding, design and construction of the sensor may have little direct connection with the data's eventual use.

A complex set of arrangements has evolved for production and dissemination of geographical data, and forms the data supply context for much of geocomputation. Recently this system has been revolutionised by the arrival of the Internet and the World Wide Web (WWW), which have removed almost entirely the costs and delays associated with traditional dissemination methods. In this new world data are to be found in widely distributed archives, ranging in size from personal servers built by individuals to make small datasets available to colleagues, to the massive servers maintained by the US Geological Survey's Eros Data Center, or the US National Aeronautics and Space Administration's EOS-DIS (Earth Observing System Data and Information System) for dissemination of vast amounts of Earth imagery and other data. To find data in this loosely coordinated and vastly complex environment the user needing data for a specific purpose must somehow:

1. *Specify* that need in terms whose meaning is widely understood.
2. Initiate a systematic process of *search*.
3. *Assess* the suitability for use of any item identified as potentially useful by the search process.
4. *Retrieve* the data using available communication channels.
5. *Open* the data for use by a local application.

This new framework differs markedly from its traditional precursor, which relied extensively on individual expertise and assistance. In most cases the potential user of data would have been a *spatially aware professional* (SAP) with knowledge of a specialised vocabulary shared with other SAPs. He or she would have interacted with a custodian of the data, perhaps at a map library or in a government office, and the telephone number of the custodian or a previous user may have been entirely sufficient to provide the necessary information about the data in question. Data would have been supplied on tape, perhaps by mailing, or in hard copy form to be digitised by the user. Much time would have been spent making the data compatible with the local application, perhaps by reformatting. Much of the available data would have been produced centrally, by a government department funded at public expense, whereas today data are increasingly available from individuals, or local agencies. With central domination of production, it was possible for uniform standards to be imposed; today, a plethora of standards have emerged as a result of marketplace competition and local autonomy.

In short, geocomputation, with its extensive data demands, is arriving as a novel paradigm at a time when many traditional arrangements for production and dis-semination of geographical data are breaking down, and are being replaced by a much more flexible, localised, autonomous, and chaotic system that is at the same time much richer, with far more to offer. While new technology has made far more data available, it has also created massive problems in making effective use of its potential. Paradoxically, only the technology itself can provide the basis of solutions. The purpose of this chapter is to examine efforts to deal with these issues, and specifically to provide tools for tackling the five stages identified above.

The next section of the chapter reviews the traditional approach to these issues, using as a framework the services provided by the research library. This is followed by five sections, one on each of the five stages of data acquisition. The chapter ends with concluding comments. Much of the discussion is based on the author's experience with the Alexandria Digital Library (Smith et al 1996), a project to make a large, distributed resource of geographical information accessible via the Internet.

4.1.1   The Library Service Model

Libraries have existed for centuries, and one of their purposes has been to satisfy the types of needs identified above, by adopting a very general approach to information retrieval. Libraries help users specify needs by providing structured tools; a thesaurus, for example, allows a user to translate terms into those accepted by the library as the basis of its own information abstraction and cataloguing. Libraries support search by abstracting information about every information object (book, journal article, or map) using standard formats. Assistance is available to the user as he or she searches for suitable objects, and assesses their fitness for use, and the user is able to browse through many information objects in searching for the best fit to a requirement because information objects are typically shelved by subject. The retrieval of information objects is made possible by assigning them unique codes. Only in the last stage, the opening of data for use by an application, is there no direct analogue among traditional library services.

While the library service model appears suited to any type of data, in practice it has not dominated the dissemination of geographical data, and alternative arrangements have emerged that are largely outside the library paradigm. Geographical data have been difficult to catalogue and abstract, cumbersome to store, and consumed by a comparatively small and specialised community. As the previous arguments and the next sections demonstrate, these assumptions are increasingly untenable, and geographical data are increasingly regarded as part of the information mainstream.

4.2   Specification

Consider the archetypical application of geocomputation. A user needs to model processes in a given geographical area, and requires data that can specify initial conditions, boundary conditions, or the variation of parameters across the area. The *footprint* of the study area is thus the most important characteristic of data, and the primary basis of search. Footprints can be defined in two ways: by specifying the bounding coordinates, or as one or more place-names. A *gazetteer* provides the ability to translate between the two options, but unfortunately few place-names have well-defined footprints, gazetteers rarely provide more than a point reference, and only certain types of place-names appear in gazetteers. Thus bounding coordinates are clearly preferable to place-names as a rigorous basis for defining both the requirements of a project and the coverage of available information objects.

Unfortunately the traditional library has no analogue of a search that is driven by a set of bounding coordinates. Information objects in the library are classified by subject, using a discrete and finite set of topics in a controlled vocabulary. Information objects are also catalogued alphabetically by author, and alphabetically by title, but in both cases the space is one-dimensional, discrete and finite. A search based on footprints is two-dimensional, continuous and infinite, and clearly not supportable using traditional library techniques, which is one reason why map libraries have been so difficult to catalogue.

Let A denote the footprint of the project, or the specification of the geographical coverage aspect of the requirement. Let B_i denote the footprint of a geographical-information-bearing object (GIBO) i. Assume that both A and B_i are defined as rectangles aligned with latitude and longitude. Although some precision is lost, the benefits of this assumption in improved performance would seem to far outweigh the disadvantages. The goodness of fit of B_i to the specification A can be measured in various ways. A simple Boolean search might require that A be wholly contained within B_i, but this would imply that all GIBOs covering the entire surface of the Earth are perfect fits to all specifications. More useful is the measure $\|A \cap B_i\|/(\|A\| \|B_i\|)^{1/2}$, or the area of intersection divided by the square root of the product of the areas. This measure is 1 if the GIBO's footprint fits the specification perfectly, and decreases if either the GIBO only covers part of the specification footprint, or the specification covers only part of the GIBO footprint. Goodchild et al (1998a) have generalised this to the case of fuzzy footprints, where the footprint of either the GIBO or the specification is uncertain or poorly defined.

After location, the specification's next most important components are likely to be theme, date and level of detail. Any geographical dataset provides information about one or more characteristics f at every location (x,y) within the footprint; theme defines the nature of f, or the dataset's *semantics*. Geographical themes range from land surface elevation to soil class, land cover class, or population density. The specification of geographical theme is complicated, however, by three issues.

First, geographical themes lack a controlled vocabulary that is comparable to those of library subject classification. There are no accepted standards for themes, and the wide range of possible themes makes it very difficult to develop one. Second, there is a tendency for GIBOs to provide information on more than one theme, either through the lumping of many *layers* into a single database, or through the assignment of multiple attributes to a single set of objects. This issue might be dealt with by changing the *granularity* of data, by breaking up a database into constituent layers and thus better-defined themes. But it is not clear that it would result in a better search process. Finally, there is the possibility that the user will define a new theme by interpretation or manipulation of the raw data, a practice that is common, for example, in the use of remotely sensed imagery or aerial photography (Curran et al, this volume). Thus a dataset that is classified as 'aerial photograph' might be used to provide information on land cover type if the user were willing to attempt an appropriate classification. In principle, therefore, every dataset should be classified by all of those themes that can be derived from the data by processing.

Theme is a discrete, nominal variable, and the goodness of fit of a GIBO's theme to the user's specification can only be measured in a binary fashion. Date is interval, however. Thus the user might specify a range of acceptable dates, and the GIBO would be identified as a 'hit' if its date fell within the range.

Any geographical dataset must have an associated *level of detail, S*. The concept of *scale* was introduced earlier with respect to the user's ability to model process, and with the implication that any dataset with a level of detail equal to or finer than *P* would be acceptable as input. Goodchild and Proctor (1997) discuss the measurement of level of detail in digital geographical datasets, and conclude that a linear measure is most appropriate, despite the massive legacy of representative fraction as a characteristic of paper maps. They also discuss the difficulties of measuring *S* for irregular geographical data models. An ideal fit to the specification would have *S=P*; in practice, however, the requirement is unlikely to be met perfectly, and instead the suitability of a GIBO with level of detail *S* against a requirement for data at scale *P* is some decreasing function $g(P-S)$ for $P>S$ (greater implies coarser or more generalised when detail and scale are expressed as linear measures). When $P<S$ the dataset may still be useful if the user has access to techniques for simulating the missing detail, and thus measuring the impact of the lack of sufficiently detailed data (Ehlschlaeger et al 1997).

In summary, the need for a geographical dataset can be specified in terms of footprint, theme, date and level of detail, plus other more specialised elements as appropriate (an exhaustive list is provided by the US Federal Geographic Data Committee's (FGDC) Content Standards for Digital Geospatial Metadata, *http://www.fgdc.gov*); and candidate information objects can be specified through *metadata* that are defined in the same terms. Goodness of fit can be measured as a binary property in some cases; as a function in the case of level of detail; and as a normalised ratio of intersection area in the case of footprints. Since there is almost no likelihood that a GIBO will match perfectly to an independently specified requirement, but a substantial likelihood that more than one dataset will be identified as potentially useful, some means must be devised for weighting these components of goodness of fit for alternative candidates, ranking the totals, and making a rational choice. Let x_{ij} denote the result of comparing GIBO$_i$ to the specification on the *j*th metadata component. Then the goodness of fit G_i will be a function $G(x_{i1}, x_{i2}, . . .)$.

4.3   Search

Armed with tools for specifying need and measuring the goodness of fit of candidate GIBOs, the discussion now turns to the process of search. There are now thousands of sites on the WWW offering GIBOs, some with restrictions on use, and some charging for use, but many offering data at no cost and without restriction. The US National Geospatial Data Clearinghouse (NGDC; *http://www.ngdc.gov/*

clearinghouse/clearinghouse.html) is one example, and many others exist in other countries, at other levels of government, and in other agencies.

Projects such as NGDC are based on the principle of *one-stop shopping*, that is, the principle that a user within some geographical domain and in need of geospatial data would go to a single, known source. All available data covering the domain would be catalogued by the source, and there would be an implicit guarantee that if the data could not be found in that source, they could not be found anywhere.

Unfortunately the one-stop shopping model is likely to fail, for several reasons. First, there is no rational basis for assigning this function to any one level in the administrative hierarchy. While it might make sense for datasets covering an entire nation to be accessible through a national server, by the same principle datasets covering an entire county should be accessible through a county server, not a national server; and what about datasets covering parts of several administrative units? The connectivity of the Internet is not perfect, and users in a given county would clearly not welcome being asked to store and access all data in one massive, global server, or the loss of control and custodianship that this would imply. Goodchild (1997) argues that the rational solution to this problem assigns each GIBO to a single server somewhere within the GIBO's footprint.

Second, projects such as NGDC are designed to serve only geographical data. While it was argued earlier that the needs of geocomputation are likely to be almost exclusively for geographical data, it does not follow that it is optimal to serve such data from exclusive servers. Goodchild (1998a) has argued that mechanisms devised for searching for geographical data can also be used to search for other types of information that are not geographical, but that nevertheless possess geographical footprints; he terms these *geographically referenced* datasets.

Finally, such projects require a high level of conformity among those who make use of them to serve data. There is a large expense in building specifications for GIBOs, particularly specifications with the richness of the FGDC metadata standard. Few incentives exist to create these specifications, other than the knowledge that by doing so one makes one's data more accessible to others. Given a choice, the custodian of data may elect to mount the data only on a small, personal server, and to provide only minimal documentation, letting the potential user bear any of the risks associated with use.

In short, any search for specified geographical data is likely to have to consider the possibility that a suitable GIBO may exist on any one of a large number of possible servers. Some means for directing the search is therefore necessary. The next sections consider two possible alternatives.

4.3.1 Search Engines

One of the most useful ways of searching for information on the WWW is to access a search engine, one of a number of sites that offer directories to the WWW. Current search engines are able to catalogue the WWW's contents by sending out intelligent agents, or *web crawlers*, to find and abstract the information available at WWW sites. They do this by following hyperlinks, or links that the custodians of sites have put in place to link to information at other sites. Information that is not linked is in a sense invisible, since web crawlers will not find it.

Web crawlers assume that information at WWW sites is in the form of text, and attempt to identify key words and phrases. The user of a search engine specifies a word or phrase, and the engine returns a list of sites determined to have information in which that word or phrase appears, in what is determined to be a significant manner (e.g. in the title of a page). Certain words and word forms are clearly more useful for this kind of search; a person's name, or a number, may be much more useful than a common word.

The catalogues produced by these search engines are very different from the metadata specifications discussed in the previous section, or the catalogues of the traditional library. There is no separation between footprint, theme, date and level of detail; instead, words and phrases must serve all purposes, and there is no guarantee that a word extracted from a body of text will in any way characterise the entire text. Moreover, GIBOs are built using the data models of GIS; if text exists in a GIBO, it does so in a very limited way in the form of attributes, or in metadata. Search engines have not been designed to abstract useful metadata from GIBOs.

Several authors have commented on the potential for a new generation of search engines that could seek out and catalogue GIBOs, generating something much closer to a metadata specification. The web crawlers associated with such a search engine would have to be able to recognise GIBOs, and to open them in order to define the key descriptors of their contents. This would clearly be much easier to do if the custodian had provided metadata in some standard format; and much easier for some geographical information formats than others. For example, a web crawler that can open a GIBO containing coordinates in some standard Earth system, such as latitude/longitude, can determine the bounding coordinates of the GIBO; but this is clearly not possible for a raster image that has no tie to the Earth's surface.

4.3.2   Collection-Level Metadata

The user of a research library has certain informal expectations about the information it is likely to contain. There is an assumption, for example, that a library at a research university will contain all important journals, and all significant books; the degree to which it does so is a commonly used measure of its success. A library will also have special collections, many of which will be unique; their existence will be known to researchers in the appropriate subject areas.

This heuristic breaks down almost completely in the digital world of the WWW. Since all users can in principle be served from a single site on the Internet, there is no need for sites to duplicate each other's contents. Instead, all WWW sites are to some degree analogous to the library's special collection, but their sheer numbers make the task of knowing which site has what virtually impossible.

In the case of GIBOs the geographical nature of the information may provide the basis for an effective heuristic that can be used to limit search. Goodchild (1997) has defined *information of geographically determined interest* (IGDI) as an information object that is of greatest interest to users in the immediate vicinity of its geographical footprint; GIBOs are a subclass of IGDIs. The servers most likely to contain a given GIBO are those closest to its footprint; and the size of the footprint also provides an indication of the level in the administrative hierarchy that is most likely to serve the GIBO. Unfortunately the architecture of the Internet makes it impossible to direct search by geographical location, although some Internet domains are geographically defined. But research efforts are currently underway to develop appropriate protocols (Navas and Imielinski 1997) that would make this much more feasible.

4.4 Fitness for Use

The assessment of fitness for use has been modelled as a comparison between the user's specification, as expressed in metadata, and the specifications of candidate datasets, with an associated metric. While some components of metadata imply a binary assessment (e.g. date), the results of other comparisons must be measured on continuous scales, leading to a ranking of candidates. Systems such as the Alexandria Digital Library (*http://alexandria.ucsb.edu*) return a number of potential candidates, ranked by a measure over which the user has some control, and limited by parameters, such as the maximum number of candidates, that are also controlled by the user.

The process of search in a library is inherently 'fuzzy' or uncertain, and it is frequently necessary for the user to *browse* in order to locate a suitable book or article. In effect, the user is unable to make a complete specification in advance, and instead refines the specification during the search process. Libraries support browsing by arranging to shelve books on similar subjects together, so that the effort on the user's part in accessing several books on the same subject is not much greater than the effort in accessing one. Similarly, a search in a digital domain should return several GIBOs, each with high score G, and allow the user to open them, browse their contents, and possibly refine the search criteria as a result. The issue of browse is discussed again later in the context of retrieval.

While any GIBO can be assessed against a specification using the methods discussed earlier, the user will also need some assurance that the GIBO's metadata are complete and accurate. It is likely that the metadata will not be complete in many cases, either because the custodian did not provide a complete specification, or because a search engine was unable to determine one. It is also possible that the data do not meet the claims made in the metadata, because of high levels of error, or because the metadata are simply incorrect.

Goodchild (1998b) has reviewed the description of data quality in metadata, and the issues involved. Goodchild et al (1998b) note that the literature on geographical data quality now includes many models, with many associated parameters, and it is increasingly unlikely that a user of geographical data would be sufficiently knowledgeable in this area to make effective use of full data quality information. Instead, they argue that data quality might be described by a *process* rather than a set of parameters. The process would be encapsulated with the data, in Java or some other code that can be executed on any client; and by initiating the process, the user would generate a number of realisations of an error model defined by the custodian or producer of the data. They argue that such a process is a full and complete specification of data quality, and yet requires no expert knowledge on the part of the user. Following Openshaw (1989), they argue that error model realisation provides a comprehensive approach to the data quality problem.

4.5 Retrieval

GIBOs have a tendency to be large; a complete representation of the US street network occupies some 10^{10} bytes, for example, and a complete Landsat scene some 300 Mb. The bandwidth available between server and client may be constrained by a modem, and will be subject to contention from other users. Thus delivery of a selected GIBO is often far from a trivial technical issue. In addition, the user may not be certain that the GIBO meets requirements until it is opened, since there will always be ambiguity and perhaps inaccuracy associated with metadata descriptions.

Server-side processing may address many of these issues, allowing the user to specify a window or other basis for selection of part of a GIBO. This is a service for which there is no obvious analogue in the traditional library, since one can only deliver part of a physical object by destroying the object, although there are many examples of server-side processing in data dissemination. It would be simple, for example, for the server to send only that subset of B_i contained in A. The degree of overlap between B_i and A has already been used in the proposed measure of 'goodness of fit'; unfortunately, cases where there are large economies to be gained by clipping B_i to A are also cases where B_i will have been given a low score because of low overlap.

More comprehensive are methods of *progressive transmission*. Suppose B_i could be organised in a hierarchical fashion, beginning with a representation of the coarsest spatial variation, and progressing to the finest details. The coarsest components would also be comparatively small in volume, and could be transmitted quickly. The user could be given the option of stopping the transmission at any point, if the GIBO appeared to be inappropriate for the requirement. Virtually any hierarchical decomposition will meet the needs of progressive transmission, including quadtrees (Samet 1990) and wavelets (Chui 1992). Unfortunately, while many suitable techniques exist for raster data, the problem of hierarchical decomposition and progressive transmission of vector data seems much more difficult, since it is essentially the problem of automated cartographic generalisation (Müller et al 1995). The viewpoint-centred methods often used in visualisation, which generalise the periphery of the field of view and thus reduce data volume, are inappropriate for geocomputation, which almost certainly requires uniform coverage of the study area.

Progressive transmission can help in two ways, by providing coarse approximations to a GIBO that the user can examine and assess, and by allowing the user to truncate transmission when some acceptable level of detail has been reached. Wavelets and other efficient decompositions have no overhead, since the hierarchically structured data occupies the same volume as the conventional form. The concept will be familiar to WWW users, since it is embedded in some image transmission standards.

4.6 Opening

Opening is the final step in the five-stage process of data acquisition for geocomputation. Having retrieved a GIBO, the user is concerned with the task of opening it in some local application, for visualisation, statistical analysis, or input to a simulation model. Note that the ability to open the GIBO has already been assumed in the previous section, where the user was able to make an informed decision based on the GIBO's contents.

It is possible to define various levels of opening. For example, the user may be able to display the form of the dataset by interpreting the coordinates defining its basic objects, but unable to interpret the attributes because no information is available defining the GIBO's semantics. Many of the details needed for successful opening, such as the name of the GIS used to create the data, may be contained in the GIBO's metadata, and transmitted as part of the *wrapper*. The Open GIS Consortium (OGC; *http://www.opengis.org*) is actively developing the standards that will allow a user application to open GIBOs of a wide range of formats and origins without any intervention on the part of the user, but it will be some time before this transparency becomes part of standard practice in geocomputation.

Opening also has no analogue among the services of the traditional library, whose responsibilities normally end when the information object is put into the hands of the user. In a digital world it is possible to imagine a host of client-side and server-side services concerned with processing information derived from distributed stores. But this suggests an awkward problem noted earlier: to what extent should metadata define not only the attributes of the GIBO, but also any form into which the GIBO can potentially be restructured or manipulated? As Kuhn (1997) argues, two datasets are essentially identical if one can be manipulated to provide the same information as the other, no matter what their actual structures may be, provided the costs of manipulation are not considered.

4.7 Conclusion

This chapter has reviewed the issues raised by a massive shift in the arrangements for retrieval of geographical information as input to geocomputation. Traditionally, the services provided by a library allowed its users to retrieve certain types of information, on the understanding that that information was to be found within physical volumes. Because digital data presented its own peculiar difficulties, early approaches to data dissemination emphasised large data centres and archives, with their own largely unique protocols. The library model did not work well for geospatial data, because of the problems of effective handling and cataloguing of maps and images, and so early efforts to adopt digital technology in support of dissemination of geospatial data occurred mostly outside the library domain.

In the past few years the growth of the WWW has opened the possibility of a generic approach to information dissemination, in which the nature of the information is largely irrelevant to the process of retrieving it. Information objects consist of 'bags of bits', with wrappers that define important characteristics of the bag's contents to the digital environment. Geographical information is thus in principle as easy to find, assess, retrieve and use as any other kind of information once it is in digital form, and many of the old distinctions based on analogue media are being questioned, reassessed, or ignored.

A five-stage process has been presented, and the chapter has discussed issues that arise at each stage, based largely on the author's experience with the Alexandria Digital Library project. While it is possible to see how each of the five stages might operate, and many efforts are under way to facilitate many of its elements, it will be many years before the legacies of previous arrangements, the problems with competing and incompatible standards, and resistance to the effort involved in providing the necessary metadata disappear, if they ever do. Until that happens, the process of search for data to support geocomputation will continue to rely, as it always has, on networks of personal contacts, idiosyncratic knowledge, luck, and the expertise of SAPs.

References

Chui C K 1992 *An introduction to wavelets*. Boston, Academic Press

Delcourt H R, Delcourt P A, Webb T III 1983 Dynamic plant ecology: the spectrum of vegetation change in space and time. *Quaternary Science Review* 1: 153

Ehleringer J R, Field C B (eds) 1993 *Scaling physiological processes: leaf to globe*. San Diego, Academic Press

Ehlschlaeger C R, Shortridge A M, Goodchild M F 1997 Visualizing spatial data uncertainty using animation. *Computers and Geosciences* 23: 387–95

George P L 1991 *Automatic mesh generation*. New York, John Wiley

Goodchild M F 1992 Geographic data modeling. *Computers and Geosciences* 18: 401–8

Goodchild M F 1997 Towards a geography of geographic information in a digital world. *Computers, Environment and Urban Systems* 21: 377–91

Goodchild M F 1998a The geolibrary. In Carver S (ed.) *Innovations in GIS 5*. London, Taylor & Francis 59–68

Goodchild M F 1998b Communicating the results of accuracy assessment: metadata, digital libraries, and assessing fitness for use. In Congalton R G, Mowrer T (eds) *Spatial accuracy assessment in natural resource analysis*. Chelsea (US), Ann Arbor Press (in press)

Goodchild M F, Proctor J D 1997 Scale in a digital geographic world. *Geographic and Environmental Modelling* 1: 5–23

Goodchild M F, Montello D R, Fohl P, Gottsegen J 1998a Fuzzy spatial queries in digital spatial data libraries. *Proceedings, IEEE–FUZZ, Anchorage, May 4–8*

Goodchild M F, Shortridge A, Fohl P 1998b Encapsulating simulation models with geospatial datasets. *Proceedings, Third International Symposium on Spatial Accuracy Assessment in Natural Resources and Environmental Sciences*

Knupp P, Steinberg S 1993 *Fundamentals of grid generation*. Boca Raton, CRC Press

Kuhn W 1997 Approaching the issue of information loss in geographic data transfers. *Geographical Systems* 4: 261–76

Müller J C, Lagrange J P, Weibel R (eds) 1995 *GIS and generalization: methodology and practice*. London, Taylor & Francis

NRC (National Research Council) 1995 *A data foundation for the national spatial data infrastructure*. Washington, DC, National Academy Press

Navas J, Imielinski T 1997 GeoCast – Geographic addressing and routing. *Proceedings, MOBICOM 97, Budapest, Hungary*: 66–76

Onsrud H J, Rushton G (eds) 1995 *Sharing geographic information*. New Brunswick, NJ, Center for Urban Policy Research

Openshaw S 1989 Learning to live with errors in spatial databases. In Goodchild M F, Gopal S (eds) *Accuracy of spatial databases*. London, Taylor & Francis: 263–76

Raper J 1997 Progress towards spatial multimedia. In Craglia M, Couclelis H (eds) *Geographic information research: bridging the Atlantic*. London, Taylor & Francis: 525–43

Rosswall T, Woodmansee G, Risser P G (eds) 1988 *Scales and global change*. New York, John Wiley

Samet H 1990 *The design and analysis of spatial data structures*. Reading (US), Addison-Wesley

Smith T R, Andresen D, Carver L, Dolin R et al 1996 A digital library for geographically referenced materials. *Computer* 29(7): 14

Part Three
DIAGNOSTICS AND PATTERN DETECTION

5

Exploratory Spatial Data Analysis in a Geocomputational Environment

Luc Anselin

Summary

This chapter reviews some methodological and technical issues associated with the implementation of exploratory spatial data analysis (ESDA) in a geocomputational environment. The emphasis is on techniques that explicitly take into account the presence of spatial autocorrelation, such as visualisation devices for spatial distributions and spatial association, local spatial association and multivariate spatial association. This is considered from both a geostatistical perspective and a lattice data perspective. Computational aspects of the integration of ESDA and GIS are reviewed both in generic terms and in the context of specific software implementations. Some ideas are outlined on potential extensions and future developments.

5.1 Introduction

Recent advances in computing hardware and software, such as object orientation, distributed computing, client–server systems, network and Internet computing have changed the nature of the spatial analyses that are demanded from geographical information systems (GIS). New concepts, such as data warehousing and spatial data mining poorly fit the paradigm of the quantitative revolution in geography of the 1960s when many of the foundations for the current collection of spatial analysis methods were established (see Goodchild, this volume; Openshaw, this volume). This quickly changing environment creates interesting challenges for the

Geocomputation: A Primer. Edited by Paul A Longley, Sue M Brooks, Rachael McDonnell and Bill Macmillan.

future of spatial data analysis as an integral part of the methodological toolbox of the geoinformation scientist.

In this chapter, I will focus on techniques of exploratory spatial data analysis (ESDA) that explicitly take into account spatial autocorrelation and spatial heterogeneity. My objective is to review and assess the current state of the art and to outline some thoughts on a number of important methodological and computational aspects of the integration of ESDA with GIS. This builds on a view of ESDA outlined in a number of other papers (e.g. Anselin 1994; 1998a) and on a framework that includes spatial data analysis as an explicit and central part in a spatial analysis module of a GIS. The latter originated in the discussion of Goodchild (1987) and Goodchild et al (1992), and extends ideas formulated earlier in Anselin and Getis (1992), Anselin et al (1993), and Anselin (1998b).

The remainder of the chapter consists of four sections. I first define the concept of ESDA and situate it relative to traditional exploratory data analysis (EDA) and cartographic visualisation. This is followed by a brief review of some common ESDA methods. Next, I consider more closely the computational aspects related to the integration of ESDA and GIS, both in generic terms as well as in the context of three specific software implementations: the ArcView/XGobi/XploRe environment of Symanzik et al (1997a; 1998a; 1998b), the S+ArcView link of MathSoft (Bao and Martin 1997), and the SpaceStat extension for ArcView (Anselin and Smirnov 1998). The chapter closes with some thoughts on potential extensions and future directions.

5.2   EDA, ESDA and Cartographic Visualisation

Exploratory spatial data analysis (ESDA) is a subset of exploratory data analysis (EDA) that focuses on the distinguishing characteristics of geographical data, and specifically on spatial autocorrelation and spatial heterogeneity (Anselin 1994; 1998a; Bailey and Gatrell 1995; Cressie 1993; Haining 1990). The basis for ESDA is the perspective towards data analysis taken in EDA. EDA consists of a collection of descriptive and graphical statistical tools intended to discover patterns in data and suggest hypotheses by imposing as little prior structure as possible (Tukey 1977). This is supposed to lead to 'potentially explicable patterns' (Good 1983: 290) and is qualitatively distinct from simple descriptive statistics. For the latter, presentation and summary of information as such are the main objectives. Modern EDA methods emphasise the interaction between human cognition and computation in the form of dynamic statistical graphics that allow the user to directly manipulate various 'views' of the data. Examples of such views are devices such as histograms, box plots, q–q plots, dot plots, and scatterplot matrices augmented with various data smoothers (for a recent review, see Cleveland 1993). State-of-the-art software implementations typically consist of dynamically linked windows in which the user is able to delete points, highlight (brush) subsets of the data, establish links between data points in different graphs, and rotate, cut through and project high-dimensional data. Specific methods include: *linked scatterplot brushing*, or ways to connect points pertaining to the same sets of observations in different scatterplots (Stuetzle 1987); *conditional plots*, consisting of a layout of panels each showing the relationship between a given set of variables (usually two) conditional on the value of one or more other variables (Becker et al 1996); *projection pursuit*, a series of static projections of a high-dimensional space, based on a criterion of optimality (Huber 1985; Friedman 1987; and the review in Cook et al 1995); and the *grand tour*, an interactive and continuous sequence of low-dimensional projections (typically in two or three dimensions) of a high-dimensional (multivariate) point cloud (Asimov 1985; and for recent reviews, Cook et al 1995; Cook and Buja 1997).

Geographical data have always played an important role as examples for illustrating dynamic graphics (e.g. in Cleveland and McGill 1988), although the importance of 'space' in these instances is typically reduced to a treatment of location as a simple x, y coordinate pair in a standard scatterplot (sometimes with a fixed graphic of polygon boundaries overlayed). The first explicit consideration of the map as a separate and integrated 'view' of the data in a dynamic graphics framework seems to be due to Monmonier (1989) who coined the term 'geographic brushing.' An early implementation of this concept is illustrated in MacDougall (1992) and by now it has been incorporated in a number of software tools for exploratory data analysis. Perhaps the most explicitly spatial of these (although conceived as separate from a GIS) is the toolbox contained in the Spider/Regard/Manet software of Haslett, Unwin and associates (Haslett et al 1990; 1991; Bradley and Haslett 1992; Unwin 1994; Haslett and Power 1995; Unwin et al 1996). In this implementation, a number of graphical and tabular statistical displays are dynamically linked, such that the selection of any subset of observations in a map or other data view is immediately reflected in all other displays. In addition to a choropleth map, typical data displays include histograms, box plots, bar charts, scatterplots and simple lists.

Ideas from the methodology of dynamic graphics in statistics have also found their way into modern practices of cartographic visualisation. While it is not always clear where cartographic visualisation ends and E(S)DA begins, several recent developments provide innovative ways to augment the static map as a device to describe spatial data (for overviews, see Hearnshaw and Unwin 1994; MacEachren and Taylor 1994; MacEachren and Kraak 1997). For example, in the cdv software of Dykes (1997) interaction with choropleth maps and cartograms includes interactive brushing, altering map symbolism, linking and probing. However, in the approach towards 'dynamic maps' in cartographic visualisation, the typical focus is on location (absolute and relative, in the sense of spatial arrangement), almost at the exclusion of multivariate attribute association (which is the main focus of traditional EDA). For example, few of the traditional statistical displays, such as box plots or scatterplots, are integrated with the map view. In contrast, true ESDA pays attention to both spatial and attribute association.

As defined by Anselin (1994; 1998a), ESDA is a collection of techniques to describe and visualise spatial distributions, identify atypical locations or spatial outliers, discover patterns of spatial association, clusters or hot spots, and suggest spatial regimes or other forms of spatial heterogeneity. Central to this conceptualisation is the notion of spatial autocorrelation or spatial association, i.e. the phenomenon where locational similarity (observations in spatial proximity) is matched by value similarity (attribute correlation) (see Cliff and Ord 1981 and Upton and Fingleton 1985 for extensive treatments). Cressie (1993) distinguishes between two data models in which spatial autocorrelation can be analysed: one based on point data as a sample of an underlying continuous distribution, or *geostatistical* data; the other consisting of a fixed collection of discrete spatial locations (points or polygons), or *lattice* data. Each requires a different set of specific methods for data exploration, although the general underlying concepts are largely the same.

5.3   ESDA Techniques

In the interest of space, this brief overview will be limited to four broad classes of techniques: visualising spatial distributions, visualising spatial association, local indicators of spatial association, and multivariate indicators of spatial association. These methods have received the most attention in the recent literature and have also been implemented in several of the software systems reviewed in Section 5.4. They are summarised in Table 5.1. The discussion will be general and references will be made to the original source materials for technical details (for more extensive reviews, see Anselin 1994; 1998a).

The tools for visualising spatial distributions are closest in spirit to many techniques of cartographic visualisation. The point of departure is slightly different. Rather than taking the map as the central element, the basis for the visualisation is a standard statistical graphics device, such as a cumulative distribution function or a box plot. Both techniques listed in the first row of Table 5.1 are based on the same principle: the visualisation of the distribution of the values of an attribute observed at a subset of locations in space. In the spatial cumulative distribution function (SCDF) introduced by Majure et al (1996), a continuous density function is estimated for a given region, based on the values at observed locations. The selection of a subset of locations (in a system with dynamically linked windows) yields highlighted points in the SCDF, and vice versa (for examples, see also Cook et al 1996; 1997 and Symanzik et al 1997a). The spatial box plot and box map achieve the same goal, but using a box plot as the device to characterise the cumulative distribution. This also leads to a natural way to identify outliers (see Anselin and Bao 1997). Again, implemented in a system with dynamically linked windows, different spatial subsets can be selected by the user, yielding different 'regional' box plots or highlighted subsets in the box plot, and vice versa (e.g. highlighting the outliers in the box plot yields their locations on the map).

The other three categories of techniques reviewed here pertain to the concept of spatial association. The main distinction between the geostatistical and lattice data methods lies in the way in which spatial proximity is formalised. In the geostatistical approach, the assumption of a continuous spatial process leads to the use of a distance metric (typically, but not necessarily, a Euclidean distance) as the means to organise observations. Since spatial association is assumed to be a smooth function

Table 5.1 *Techniques of ESDA*

	Geostatistical perspective	Lattice perspective
Visualising spatial distribution	• spatial cumulative distribution function	• box map • regional histograms • spatial exploratory analysis of variance
Visualising spatial association	• spatially lagged scatterplot • variogram cloud plot • variogram box plot	• spatial lag charts • Moran scatterplot and map
Local spatial association	• outliers in variogram box plot • outliers in variogram cloud plot	• LISA maps • outliers in Moran scatterplot
Multivariate spatial association	• multivariate variogram cloud plot	• multivariate Moran scatterplot

of distance, a formal measure of value (dis)similarity between two observations, such as the squared difference, is compared to the distance that separates them. A high degree of spatial autocorrelation implies small differences at close distances and increasing differences at higher distances. The formal function that operationalises this notion is the variogram (see, for example, Cressie 1993 for technical details). ESDA techniques from the geostatistical realm consist of ways in which the variogram can be visualised, summarised and probed for the presence of local non-stationarities or other non-standard behaviour. An important aspect of this visualisation is that the entities in a variogram pertain to *pairs* of observations (separated by a given distance), and not to the individual locations.

In the lattice approach towards spatial association, the discrete nature of the observations (a countable number of points or polygons in space) suggests a different way to formally express the coincidence of locational and value similarity. The core concept here is the notion of spatial neighbour, which leads to the construction of *spatial weights* matrices and spatially lagged variables. A spatial weights matrix contains a row for each observation in which the non-zero elements (typically equal to one) stand for the neighbours. Such neighbours can be defined in a number of ways, based on contiguity (common boundaries) or distance criteria (within a given critical distance of each other). For each location, the value for an attribute at that location can be compared to the value at neighbouring locations, typically summarised in the form of a weighted average of the observations at those locations, or a *spatial lag*. Locations where the attribute is similar to the average for the neighbours suggest positive spatial association, whereas the opposite (a high value surrounded by a low average for the neighbours, or low value surrounded by a high average for neighbours) suggests negative spatial association, or a spatial outlier (see Anselin 1995a; 1996 for extensive discussions and illustrations). ESDA tools based on the lattice data approach therefore are based on a visualisation of the association between variables and their spatial lags, for different definitions of value similarity (e.g. cross-product, squared difference) and for different spatial weights (first and higher orders of contiguity, different distance bands).

The techniques in the second row of Table 5.1 pertain to the visualisation of global spatial association. In the geostatistical approach, this is accomplished by means of a spatially lagged scatterplot (Cressie 1984), variogram cloud plot and variogram box plot (Haslett et al 1991; see also Curran et al, this volume). In the lattice data approach, the respective tools are the spatial lag chart (Anselin and Bao 1997) and the Moran scatterplot/map (Anselin 1994; 1996). The principle behind the spatially lagged scatterplot and the spatial lag chart is very similar: in the former, the values at locations within a given distance band are plotted against each other (a 45 degree linear association indicating strong spatial autocorrelation), while in the latter the value at each location is charted (bar chart, pie chart) against the spatial lag (equal height bars indicate positive spatial autocorrelation, unequal heights suggest spatial outliers). A variogram cloud plot and associated box plot consists of all the squared differences (or robust equivalents: see Cressie and Hawkins 1980) plotted against the distance bands to which they pertain. The mean or median tendency for each distance band suggests an overall pattern for the change in association with distance (typically, greater differences for greater distances). In a Moran scatterplot, the spatial lag (weighted average of values at neighbouring locations) is plotted against the value at each location and the slope of the regression line through this scatter corresponds to the familiar Moran's I statistics for spatial autocorrelation. The steeper the slope, the stronger is the degree of autocorrelation. In addition, the scatterplot naturally divides the type of spatial association into four categories, two for positive spatial autocorrelation (high values surrounded by high values, low values surrounded by low values) and two for negative spatial autocorrelation (high values surrounded by low values and vice versa). These four categories result in a form of spatial smoothing that can be visualised on a map in a straightforward manner (e.g. Anselin and Bao 1997).

Visualisation of local spatial autocorrelation can be accomplished by focusing on the outliers in the plots described above. For example, identifying the points in a variogram box plot that are outside the 'fences' can suggest which pairs of locations have a value similarity that is distinct from that of the other pairs in the corresponding distance band. In a dynamic graphics framework such pairs of observations can be connected on a map by means of a line (e.g. Symanzik et al 1997a). In contrast, outliers in a Moran scatterplot correspond to specific locations (rather than pairs) and can be identified by means of the standard regression diagnostics (see Anselin 1995a; 1996 for technical details). An additional tool for dealing with local spatial association in the lattice approach is the LISA map, or map depicting locations with significant statistics of local spatial association (Getis and Ord 1992; Anselin 1995a; Ord and Getis 1995; Unwin 1996). Such locations suggest the presence of hot spots or spatial outliers and can be readily computed for a range of spatial contiguity weights or distance bands.

Techniques for the exploration of multivariate spatial association are currently still in their infancy. They are based on a generalisation of the variogram in multiple dimensions to the cross-variogram (Ver Hoef and Cressie 1993) or on a similar extension of Moran's *I* statistic of spatial autocorrelation (Wartenberg 1985). The former yields a multivariate lagged scatterplot and variogram cloud plot (Majure and Cressie 1997), while the latter is implemented as a straightforward generalisation of the Moran scatterplot (Anselin and Smirnov 1998). Again, as an element in a dynamic graphics framework, these devices allow for the interactive identification of outliers and other interesting locations.

Some of these ideas are illustrated in Plate II, an application of the SpaceStat extension for ArcView (Anselin and Smirnov, 1998), which is discussed in more detail in Section 5.4.2.3. The map in View 1 illustrates the spatial distribution of crime in neighbourhoods in Columbus, Ohio in the form of a choropleth map, with the value of crime and its spatial lag superimposed on each neighbourhood centroid as a bar chart (a so-called spatial lag bar chart). The spatial association in the data is further visualised by means of a Moran scatterplot map (Moran Scatter 2 in the View), which symbolises the four quadrants of the Moran scatterplot (a linked chart). The red values near the centre of the city indicate high-crime neighbourhoods (above average crime rate) that are surrounded by high-crime neighbourhoods (above average spatial lag). They correspond to the upper right hand quadrant in the scatterplot. The pink values in the suburbs indicate low–low clusters. The green neighbourhoods are spatial 'outliers', low crime rates surrounded by high values (bright green) and vice versa (dark green). Two additional maps show the locations with a significant local indicator of spatial association, respectively a local Moran (View 1 – LISA Map 1) and a G_i statistic (View 1 – GStat Map 1). Parenthetically, this also suggests a significant cluster of crime rates near the centre of the city. Further insight into the pattern of spatial association is provided by the dynamically linked graphs, which are invoked by means of the DynESDA floating toolbar. The Moran scatterplot shows the global Moran statistic as the slope of the scatterplot (significance is assessed by means of a randomization approach, not illustrated in Plate III). The scatterplot is linked with the LISA map as well as with a box plot of individual LISA statistics (Local Moran). The yellow points on the scatterplot were highlighted by means of a select or brushing tool and linked to their corresponding values in the Local Moran box plot and LISA map. One of these low crime areas corresponds to a spatial outlier (upper left quadrant in the scatterplot), which is also an outlier in the box plot (very lowest yellow point) and a significant location in the LISA map (the leftmost yellow neighbourhood was also 'red', or significant at 0.001). This may suggest that the corresponding location does not fit the overall pattern. In contrast, the other selected point is central in the Local Moran box plot and does not correspond with a significant LISA. While this static view only provides a limited insight into the capabilities of a dynamically linked framework, it may suggest some of the potential of such an approach.

5.4   Integration of ESDA and GIS

The integration of spatial analysis within GIS has received considerable attention in both the academic and commercial world since Goodchild (1987) suggested its central role as part of the development of a geographic information science (see also Goodchild 1992). A large number of conceptual frameworks have been outlined, dealing both with the types of analyses that should be included as part of the spatial analysis 'toolbox' in a GIS, as well as with the ways in which linkages between specific software packages may be implemented (see, among many others, Anselin 1998b; Openshaw 1991; Anselin and Getis 1992; Goodchild et al 1992; Anselin et al 1993; Griffith 1993; Bailey 1994; Haining 1994; Openshaw and Fischer 1995; and the collections of papers in Fotheringham and Rogerson 1994; Fischer et al 1996; Fischer and Getis 1997).

Early implementations of these ideas focused on establishing links between commercial GIS software and statistical packages. In the terminology of Goodchild et al (1992), these forms of integration were mostly based on *loose coupling*, i.e. data and commands were passed back and forth between the packages by means of auxiliary files (early examples include Farley et al 1990 and Flowerdew and Green 1991). *Close coupling*, i.e. a framework where commands in one software system are callable from another system by means of seamless inter-process communication, has only been implemented more recently, e.g. in the S+Gislink between Arc/Info and the S-Plus statistical system (MathSoft 1996a). However, these implementations of software integration mostly consist of augmenting the analysis capabilities in a GIS with the range of standard statistical procedures available in a commercial statistical software package, but with a few rare exceptions they are not focused on *spatial* data analysis methods.

The addition of ESDA to the GIS analysis toolbox has been pursued in a number of different ways. One approach, which could be termed *encompassing* (using the terminology of Anselin and Getis 1992), consists of writing spatial data analysis routines in a system's macro or scripting language (e.g. AML for Arc/Info or Avenue for ArcView). Such an approach is fully integrated within the GIS interface and hides the linked nature of the spatial data routines from the user. Some examples of this are the routines for computing global and local spatial auto-correlation indices in Arc/Info (e.g. Ding and Fotheringham 1992; Bao et al 1995; Can 1996) and in ArcView (Zhang and Griffith 1997). However, these imple-mentations are typically slow in execution, due to the deficiencies of the scripting language in terms of algorithms and data structures necessary to carry out statistical computations. Moreover, the size of datasets that can be analysed with these approaches is severely limited and precludes realistic applications (see the discussion in Anselin and Bao 1997). A different, *modular* approach consists of including spatial data analysis software (typically specially developed routines) in a collection of linked systems where the communication between the different entities is established by means of a combination of loose and close coupling. Recent examples of this approach are the combination of Arc/Info with SpaceStat, XGobi and clustering software in Zhang et al (1994), the close coupling between ArcView, XGobi and most recently XploRe in the work at the Iowa State University Statistics Laboratory (e.g. Symanzik et al 1994; 1996; 1997a; 1998a; 1998b), the integration between Arc/Info, spatial statistical analysis and regionalisation routines in the SAGE system developed at Sheffield University (Haining et al 1996; Wise et al 1997), and the linkage between ArcView and SpaceStat in Anselin and Bao (1997).

Several computational issues must be addressed in the addition of ESDA to a GIS environment. I will next review some of these in the context of a generic *dynamically linked spatial association visualiser*. This is followed by a brief assessment of three currently operational implementations of such a device.

5.4.1   A Dynamically Linked Spatial Association Visualiser

Visualising spatial association is a central component of both geostatistical and lattice data approaches towards ESDA. However, the implementation of a device that accomplishes this is not straightforward and requires a number of specialised data structures and linking mechanisms.

Figure 5.1 illustrates the necessary data structures. They are defined in general terms and could easily be embedded into an object-oriented framework by encompassing them into classes and objects (e.g. as in the Avenue scripting language for ArcView). The observations are the point of departure, shown on the left hand side of the figure. They are conceptualised as objects in both locational and attribute space. Location objects are characterised by a unique identifier (i) and either point coordinates (x_i, y_i) or polygon boundaries (x_{i1}, y_{i1}; x_{i2}, y_{i2}, . . ., x_{ik}, y_{ik}), while attributes are contained in a table (z_i). Both location and attribute information are linked by means of the unique identifier in a standard way (e.g. as implemented in a relational database). Note that the distinction between points and polygons is not crucial, since common spatial transformation functions allow points to be converted to polygons (tessellation), or polygons to be converted to points (centroids). The locational information is typically sufficient to allow most mapping functions. However, in order to compute the necessary spatial association statistics, auxiliary spatial data structures must be derived. These are given on the right hand side of Figure 5.1. For the variogram, they are all interpoint distances (as well as angles, for non-isotropic models); for the Moran scatterplot, they are the topology or contiguity structure of the polygons (many GIS contain an arc-node data structure that facilitates the construction of a spatial weights matrix, although such a weights matrix itself is typically not part of the GIS). For large datasets, the efficient computation and storage of distance and contiguity information is not trivial and should take into account sparsity and spatial structure. For simplicity's sake, the data structures are often referred to as spatial weights and distance matrices, although a matrix is not an efficient way to manipulate this information (see Anselin and Smirnov 1996; Smirnov 1998).

In addition to interpoint distances, the variogram requires the computation of a new data item for each pair of observations, i.e. the squared differences or $z_{ij} = (z_i - z_j)^2$. There are $O(n^2)$ of such squared differences and they require a specialised data structure as well (in principle, the matrix of squared differences can be added to a typical data table, but this is not efficient except for trivial applications). The Moran scatterplot requires the computation of the spatial lag for each observation, or $Wz_i = \sum_j w_{ij} z_j$ (where w_{ij} are the elements of the spatial weights matrix). In contrast to the variogram elements, the spatial lags can be added to the usual attribute table in standard fashion.

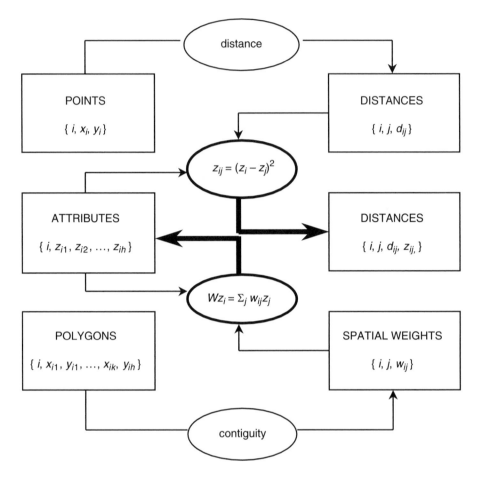

Figure 5.1   *Data structures for the spatial association visualiser*

The visualisation of spatial association is carried out by means of two specialised scatterplots, the Moran scatterplot and the variogram cloud plot. The latter is constructed by taking the distance d_{ij} as the x-axis and squared differences z_{ij} as the y-axis, while the former plots the value of the z_i on the x-axis and the matching spatial lag Wz_i on the y-axis. The dynamic linkage between these objects and the map or other statistical graphics differs, since the elements of a Moran scatterplot pertain to single observations, while those of the variogram cloud plot pertain to pairs of observations. The implementation of the linkage should be such that a request addressed to an element in each object is replicated for the matching elements in the linked objects. For example, the link between the Moran scatterplot object and the map should be such that a select or deselect request addressed to a point in the plot or polygon on the map should trigger the same request in the matching elements in the other object. This is fairly straightforward to implement. In contrast, in order to obtain dynamic linking between the variogram cloud plot and the map, an additional translation must be carried out, which identifies the origin and destination (the *i* and *j*) for each squared difference, or, alternatively, finds each pair in which a selected location (on the map) is either *i* or *j*. This is typically visualised by lines connecting these points (rather than by individually highlighted points) and requires a specialised linking mechanism. Moreover, a direct linkage between points in the variogram cloud plot and pairs of observations in other statistical graphics (such as histograms and box plots) is not useful. Therefore, two separate linking mechanism must be supported (one for single observations, one for pairs).

5.4.2   Practical Implementations

Aspects of this generic spatial association visualiser are contained in a number of recently developed implementations of linked ESDA and GIS. Examples are the ArcView/XGobi/XploRe link (Symanzik et al 1997a), the S+ArcView link (Bao and Martin 1997) and the SpaceStat extension for ArcView (Anselin and Smirnov 1998). In each implementation, the linking mechanism is approached slightly differently.

5.4.2.1   *ArcView/XGobi/XploRe*

The integration of the ArcView GIS (ArcView 3.0; ESRI 1996) with the high-dimensional data visualisation software XGobi (Buja et al 1996) and the regression exploration software XploRe (Härdle et al 1995) is a form of close coupling where commands from one system are invoked by another by means of remote procedure calls in a client–server architecture (RPC on UNIX platforms). In this set-up, each system can function as both a server and a client, although there can only be a single linkage open at any given time (Symanzik et al 1998a). The main focus in this framework is on geostatistical approaches, and only point data structures are supported at the time of writing. The variogram cloud plot is operationalised as a scatterplot in XGobi, based on the *x,y* coordinates and attribute data that are passed from ArcView. A special data structure, called an *array of pairs* is used to support a dynamic link between points in a View in ArcView and the squared difference points in the variogram cloud plot. This matrix contains the two identifiers for each pair used in the variogram.

   Brushing in this linked framework is controlled by XGobi. The selection of a point or set of points in the variogram cloud plot is passed through the array of pairs data structure to find the identifiers for each pair. This information is then passed back to ArcView to trigger a line to be drawn between the matching points in the View. Brushing from within ArcView (i.e. the 'selection' of specific points) is implemented indirectly: the selection is first passed to XGobi (without any action in the View), where the appropriate pairs are identified and highlighted in the scatterplot. In a second step, this information is then passed back to ArcView for line drawing, similar to the one-way XGobi–ArcView link. In practice, the size of datasets that can be analysed in this fashion is limited by the use of the array of pairs data structure. The development of efficient algorithms for the computation and storage of distance information becomes crucial in this respect.

5.4.2.2 *S+ArcView Link*

The S+ArcView link (Bao and Martin 1997) is at the time of writing the only implementation of an integrated ESDA with GIS in a commercial environment. The distinguishing characteristic of this linkage is a close coupling between the S-Plus software (S-plus 4.0, MathSoft 1997) as a statistical analysis server and ArcView as the client. This is obtained through a set of specialised functions that take advantage of the S+API (applications programming interface) and are loaded as an 'extension' to ArcView. This allows users to enter S-Plus commands through a graphical user interface in ArcView and to pass spatial data (centroid coordinates, distances, contiguity weights) to S-Plus for use in the S+Spatialstats module (MathSoft 1996b), in which a variogram cloud plot is readily implemented. A Moran scatterplot can be executed in S-Plus as a user-defined function. Both point and polygon data objects are supported.

The link between S-Plus and ArcView is slightly different, in the sense that data contained in S-Plus objects are written to an intermediate file before they are linked to a table in ArcView. This form of loose coupling is currently not fully reliable since a built-in mechanism for a consistent and unique identifier of each observation in both systems is lacking and this becomes the user's responsibility. As a result, true dynamic linking between graphs in S-Plus and Views in ArcView is currently not possible. On the other hand, the tools in S-Plus are flexible enough to provide a crude form of linking, through loose coupling. For example, with the spatial weights matrix obtained from ArcView, a Moran Scatterplot can be constructed in S-Plus as a graph. Selected points on this graph can be 'identified' and their identifiers (observation numbers) passed back to ArcView in a file that can be joined with the current attribute table. A simple query can highlight the selected observations in a view (e.g. by querying for non-zero values of a selection dummy variable). It should be noted that the S+ArcView linkage is still prototypical at this point in time and in principle it would be possible to establish a two-way interaction between the software systems, in which ArcView could be server as well as client. However, some features of the current underlying design of S (and S-Plus) may preclude a true dynamic interaction between S-Plus objects and other software systems at this point.

5.4.2.3 *SpaceStat Extension for ArcView*

The SpaceStat extension for ArcView (Anselin and Smirnov 1998) is a further development of the linkage between the SpaceStat specialised software for spatial data analysis (Anselin 1995b) and a GIS such as ArcView, originally outlined in Anselin et al (1993) and Anselin and Bao (1997). While the latter is based on loose coupling and data transfer by means of auxiliary files, the current SpaceStat extension implements a form of close coupling. The extension consists of a collection of dynamically linked libraries (DLL) written in C++ that are called from

Avenue scripts within ArcView. The libraries contain the basic functionality of a dynamic ESDA for lattice data in the form of histograms, box plots, scatterplots and a Moran scatterplot. They are invoked as objects from within an Avenue script. Several objects may be open at any one time and they are fully dynamically linked among themselves. In the current implementation, the linkage is only effective with respect to a single view, in the sense that brushed items on any of the statistical graphs will be highlighted as selected in only a single view within ArcView. Other views are unaffected and cannot interact with the graphs. While the functionality of the extension is still fairly limited, it was developed from scratch and based on efficient algorithms and data structures to perform ESDA. Thereby some of the compromises that must be made when linking existing software systems were avoided, thus resulting in better performance.

5.5   Potential Extensions and Future Directions

In a recent paper, Buja et al (1996) suggest a taxonomy for interactive high-dimensional data visualisation consisting of three data analytic tasks and three types of interactive view manipulation. The tasks considered are finding Gestalt (or, locating features of interest in the data), posing queries (finding interesting patterns), and making comparisons (relating variables or projections to each other or comparing subsets of the data). Three categories of tools are suggested to support these data analytic tasks: focusing individual views (e.g. selecting variables, projections or subsets of the data), linking multiple views (brushing), and arranging many views (e.g. scatterplot matrices): for details, see Buja et al (1996: 79–81). Many aspects of ESDA easily fit into this taxonomy, such as the spatial density functions and linked windows. Their properties are thus well known and can be derived as special cases of a general framework. However, other ESDA concepts do not readily fit the existing taxonomy. For example, the map is conspicuously absent as a separate view of data (while time series plots are included), and many of the standard manipulations pertain to projections and rotations in variable space, but do not easily extend to locational space. Hence, further work is needed to integrate the 'spatial' in ESDA more firmly in the established traditions and methods of EDA and to develop specialised techniques where this cannot be accomplished.

Advances in the current state of the art in ESDA will likely come from the transformation of existing tools for high-dimensional data visualisation into devices that more explicitly take into account location and spatial association. Some of these extensions are readily obtained, such as the combination of brushing a Moran scatterplot as part of a matrix of regular (i.e. non-spatial) scatterplots, potentially linked with histograms and choropleth maps for multiple variables. This can be considered as a form of conditioning in the sense of Becker and Cleveland (1987) and is in fact already contained in some of the software systems discussed in Section 5.4. However, the application of methods such as projection pursuit and the grand tour to higher dimensional point clouds that include relative location (i.e. contiguity and spatial arrangement) is not immediately obvious. Similarly, techniques developed for the exploration of time series data, even multiple time series, do not directly generalise to space–time data. A major methodological roadblock in this respect is the need to properly define a metric for space–time 'contiguity', i.e. to define which combinations of locations and time periods are 'neighbours' to a given observation in space–time. This necessitates a careful consideration of the dynamics of the processes under consideration. Insight into these dynamics may be gained by the clever application of animation techniques and other emerging visualisation tools. In addition, new data models and data structures may be necessary to add notions of 'contiguity' to the standard locational information in an efficient manner so that effective algorithms for the computation of higher order contiguity and distance-based weights may be implemented (see Anselin and Smirnov 1996; Smirnov 1998).

Most current techniques of exploratory data analysis work fine for small- to medium-sized datasets (up to a few thousands of observations). However, increasingly large spatial datasets become the subject of investigation in spatial analysis (with 100 000 to millions of observations), primarily in applications of remote sensing (Curran et al, this volume), but also in geodemographics and marketing analysis. Such datasets strain both methodology and computational resources. For example, as illustrated in Wegman (1995: 287), statistical operations of $O(n^2)$ are unrealistic in currently existing computing environments, since they require the resources of architectures with teraflop capabilities, even to manipulate large datasets (10^8 observations). Interactive operations, which are essential for high-dimensional data visualisation and GIS operations encounter the same constraint much sooner, for medium-sized datasets (10^6 observations). Many methods of ESDA, especially those based on the geostatistical approach, require the consideration (storage) of n^2 data points for the computation of variables such as distances and contributions to the variogram. Simple extrapolation of these methods to very large datasets is therefore not feasible. New approaches are needed that use efficient algorithms, implement sparsity and possibly focus on subsets of the data based on careful spatial sampling. Ideas from the LISA literature may be useful as a point of departure for embedding explicit notions of location and spatial association in the search for clusters and other data mining approaches, which is currently not the case (see, for example, Murray and Estivill-Castro 1998). Future developments in the integration of the GIS functions of storage and query for very large datasets may exploit such notions of local spatial association. It is clear that methodological developments cannot be considered in isolation from their (geo)computational implications. Effective integration of ESDA within GIS must involve both aspects.

References

Anselin L 1994 Exploratory spatial data analysis and geographic information systems. In Painho M (ed.) *New tools for spatial analysis*. Luxembourg, Eurostat: 45–54

Anselin L 1995a Local indicators of spatial association – LISA. *Geographical Analysis* 27: 93–115

Anselin L 1995b *SpaceStat Version 1.80*. Morgantown, Regional Research Institute, West Virginia University

Anselin L 1996 The Moran scatterplot as an ESDA tool to assess local instability in spatial association. In Fischer M, Scholten H, Unwin D (eds) *Spatial analytical perspectives on GIS*. London, Taylor & Francis

Anselin L 1998a Interactive techniques and exploratory spatial data analysis. In Longley P A, Goodchild M F, Maguire D J, Rhind D W (eds) *Geographical information systems: principles, techniques, management and applications*. New York: John Wiley: 1: 251–64

Anselin L 1998b GIS research infrastructure for spatial analysis of real estate markets. *Journal of Housing Research* 9 (in press)

Anselin L, Bao S 1997 Exploratory spatial data analysis linking SpaceStat and ArcView. In Fisher M, Getis A (eds) *Recent developments in spatial analysis*. Berlin, Springer: 35–59

Anselin L, Getis A 1992 Spatial statistical analysis and geographic information systems. *Annals of Regional Science* 26: 19–33

Anselin L, Smirnov O 1996 Efficient algorithms for constructing proper higher order spatial lag operators. *Journal of Regional Science* 36: 67–89

Anselin L, Smirnov O 1998 *The SpaceStat extension for ArcView 3.0*. Morgantown, Regional Research Institute, West Virginia University

Anselin L, Dodson R, Hudak S 1993 Linking GIS and spatial data analysis in practice. *Geographical Systems* 1: 3–23

Asimov D 1985 The grand tour: a tool for viewing multidimensional data. *SIAM Journal on Scientific and Statistical Computing* 6: 128–43

Bailey T C 1994 A review of statistical spatial analysis in geographical information systems. In Fotheringham S, Rogerson P (eds) *Spatial analysis and GIS*. London, Taylor & Francis: 13–44

Bailey T C, Gatrell A 1995 *Interactive spatial data analysis*. Harlow, Longman

Bao S, Martin D 1997 *User's reference for the S+ArcView link*. Seattle, MathSoft Inc, Data Analysis Products Division

Bao S, Henry M, Barkley D, Brooks K 1995 RAS: a regional analysis system integrated with Arc/Info. *Computers, Environment and Urban Systems* 18: 37–56

Becker R, Cleveland W S 1987 Brushing scatterplots. *Technometrics* 29: 127–42

Becker R, Cleveland W S, Shyu M-J 1996 The visual design and control of Trellis display. *Journal of Computational and Graphical Statistics* 5: 123–55

Bradley R, Haslett J 1992 High interaction diagnostics for geostatistical models of spatially referenced data. *The Statistician* 41: 371–80

Buja A, Cook D, Swayne D F 1996 Interactive high-dimensional data visualization. *Journal of Computational and Graphical Statistics* 5: 78–99

Can A 1996 Weight matrices and spatial autocorrelation statistics using a topological vector data model. *International Journal of Geographical Information Systems* 10: 1009–17

Cleveland W S 1993 *Visualizing data*. Summit, NJ, Hobart Press

Cleveland W S, McGill M E 1988 *Dynamic graphics for statistics*. Pacific Grove, Wadsworth

Cliff A, Ord J K 1981 *Spatial processes: models and applications*. London, Pion

Cook D, Buja A 1997 Manual controls for high-dimensional data projections. *Journal of Computational and Graphical Statistics* 6: 464–80

Cook D, Buja A, Cabrera J, Hurley C 1995 Grand tour and projection pursuit. *Journal of Computational and Graphical Statistics* 4: 155–72

Cook D, Majure J J, Symanzik J, Cressie N 1996 Dynamic graphics in a GIS: exploring and analyzing multivariate spatial data using linked software. *Computational Statistics* 11: 467–80

Cook D, Symanzik J, Majure J J, Cressie N 1997 Dynamic graphics in a GIS: more examples using linked software. *Computers and Geosciences* 23: 371–85

Cressie N 1984 Towards resistant geostatistics. In Verly G, David M, Journel A G, Marechal A (eds) *Geostatistics for natural resources characterization* (Part 1). Dordrecht, Reidel: 21–44

Cressie N 1993 *Statistics for spatial data*, revised edition. New York, John Wiley

Cressie N, Hawkins D M 1980 Robust estimation of the variogram, I. *Journal of the International Association for Mathematical Geology* 12: 115–25

Ding Y, Fotheringham A S 1992 The integration of spatial analysis and GIS. *Computers, Environment and Urban Systems* 16: 3–19

Dykes J A 1997 Exploring spatial data representation with dynamic graphics. *Computers and Geosciences* 23: 345–70

ESRI 1996 *ArcView 3.0*. Redlands, Environmental Systems Research Institute

Farley J A, Limp W F, Lockhart J 1990 The archaeologist's workbench: integrating GIS, remote sensing, EDA and database management. In Allen K, Green F, Zubrow E (eds) *Interpreting space: GIS and archaeology*. London, Taylor & Francis: 141–64

Fischer M M, Getis A 1997 *Recent developments in spatial analysis*. Berlin, Springer

Fischer M M, Scholten H J, Unwin D 1996 *Spatial analytical perspectives on GIS*. London: Taylor & Francis

Flowerdew R, Green M 1991 Data integration: statistical methods for transferring data between zonal systems. In Masser I, Blakemore M (eds) *Handling geographical information*. London, Longman: 38–54

Fotheringham S, Rogerson P 1994 *Spatial analysis and GIS*. London, Taylor & Francis

Friedman J H 1987 Exploratory projection pursuit. *Journal of the American Statistical Association* 82: 249–66

Getis A, Ord J K 1992 The analysis of spatial association by use of distance statistics. *Geographical Analysis* 24: 189–206

Good I J 1983 The philosophy of exploratory data analysis. *Philosophy of Science* 50: 283–95

Goodchild M F 1987 A spatial analytical perspective on geographical information systems. *International Journal of Geographical Information Systems* 1: 327–34

Goodchild M F 1992 Geographical information science. *International Journal of Geographical Information Systems* 6: 31–45

Goodchild M F, Haining R P, Wise S et al 1992 Integrating GIS and spatial analysis – problems and possibilities. *International Journal of Geographical Information Systems* 6: 407–23

Griffith D A 1993 Which spatial statistical techniques should be converted to GIS functions? In Fischer M M, Nijkamp P (eds) *Geographic information systems, spatial modeling and policy evaluation*. Berlin, Springer: 101–14

Haining R F 1990 *Spatial data analysis in the social and environmental sciences*. Cambridge, Cambridge University Press

Haining R F 1994 Designing spatial data analysis modules for geographical information systems. In Fotheringham S, Rogerson P (eds) *Spatial analysis and GIS*. London, Taylor & Francis: 45–63

Haining R F, Ma J, Wise S 1996 Design of a software system for interactive spatial statistical analysis linked to a GIS. *Computational Statistics* 11: 449–66

Härdle W, Klinke S, Turlach B A 1995 *XploRe: an interactive statistical computing environment*. Berlin, Springer

Haslett J, Power G M 1995 Interactive computer graphics for a more open exploration of stream sediment geochemical data. *Computers and Geosciences* 21: 77–87

Haslett J, Wills G, Unwin A 1990 Spider, an interactive statistical tool for the analysis of spatially distributed data. *International Journal of Geographical Information Systems* 4: 285–96

Haslett J, Bradley R, Craig P, Unwin A, Wills G 1991 Dynamic graphics for exploring spatial data with applications to locating global and local anomalies. *The American Statistician* 45: 234–42

Hearnshaw H M, Unwin D J 1994 *Visualisation in geographical information systems*. Chichester, John Wiley

Huber P J 1985 Projection pursuit. *The Annals of Statistics* 13: 435–525

MacDougall E B 1992 Exploratory analysis, dynamic statistical visualization, and geographic information systems. *Cartography and Geographic Information Systems* 19: 237–46

MacEachren A M, Kraak M-J 1997 Exploratory cartographic visualization: advancing the agenda. *Computers and Geosciences* 23: 335–43

MacEachren A M, Taylor D R F 1994 *Visualization in modern cartography*. Oxford, Pergamon

Majure J, Cressie N 1997 Dynamic graphics for exploring spatial dependence in multivariate spatial data. *Geographical Systems* 4(2): 131–58

Majure J, Cook D, Cressie N, Kaiser M, Lahiri S, Symanzik J 1996 Spatial CDF estimation and visualization with applications to forest health monitoring. *Computing Science and Statistics* 27: 93–101

MathSoft 1996a *S+Gislink*. Seattle, MathSoft Inc Data Analysis Products Division

MathSoft 1996b *S+Spatialstats*. Seattle, MathSoft Inc Data Analysis Products Division

MathSoft 1997 *S-Plus 4.0*. Seattle, MathSoft Inc Data Analysis Products Division

Monmonier M 1989 Geographic brushing: enhancing exploratory analysis of the scatterplot matrix. *Geographical Analysis* 21: 81–84

Murray A, Estivill-Castro V 1998 Cluster discovery techniques for exploratory spatial data analysis. *International Journal of Geographical Information Systems* 12 (in press)

Openshaw S 1991 Developing appropriate spatial analysis methods for GIS. In Maguire D J, Goodchild M F, Rhind D W (eds) *Geographical information systems: principles and applications*, Vol. 1. Harlow, Longman: 389–402

Openshaw S, Fischer M M 1995 A framework for research on spatial analysis relevant to geo-statistical information systems in Europe. *Geographical Systems* 2: 325–37

Ord J K, Getis A 1995 Local spatial autocorrelation statistics: distributional issues and applications. *Geographical Analysis* 27: 286–306

Smirnov O 1998 *Computational aspects of spatial data analysis*. Unpublished PhD dissertation, Division of Resource Management, West Virginia University, Morgantown

Stuetzle W 1987 Plot windows. *Journal of the American Statistical Association* 82: 466–75

Symanzik J, Majure J, Cook D, Cressie N 1994 Dynamic graphics in GIS: a link between Arc/Info and XGobi. *Computing Science and Statistics* 26: 431–5

Symanzik J, Majure J, Cook D 1996 Dynamic graphics in a GIS: a bidirectional link between ArcView 2.0 and XGobi. *Computing Science and Statistics* 27: 299–303

Symanzik J, Majure J J, Cook D, Megretskaia I 1997a Linking ArcView 3.0 and XGobi: insight behind the front end. Preprint 97–10, Department of Statistics, Iowa State University, Ames

Symanzik J, Megretskaia I, Majure J J, Cook D 1997b Implementation issues of variogram cloud plots and spatially lagged scatterplots in the linked ArcView 2.1 and XGobi environment. *Computing Science and Statistics* 28: 369–74

Symanzik J, Klinke S, Schmelzer S, Cook D 1998a The ArcView/XGobi/XploRe environment: technical details and applications for spatial data analysis. *ASA Proceedings of the Section on Statistical Graphics*. Alexandria, American Statistical Association (in press)

Symanzik J, Kötter T, Schmelzer S, Klinke S, Cooke D, Swayne D 1998b Spatial data analysis in the dynamically linked ArcView/XGobi/XploRe environment. *Computing Science and Statistics* 29 (in press)

Tukey J W 1977 *Exploratory data analysis*. Reading, Addison-Wesley

Unwin A 1994 REGARDing geographic data. In Dirschedl P, Osterman R (eds) *Computational statistics*. Heidelberg, Physica: 345–54

Unwin A 1996 Exploratory spatial analysis and local statistics. *Computational Statistics* 11: 387–400

Unwin A, Hawkins G, Hofman H, Siegl B 1996 Interactive graphics for data sets with missing values – MANET. *Journal of Computational and Graphical Statistics* 5: 113–22

Upton G J, Fingleton B 1985 *Spatial data analysis by example*. New York, John Wiley

Ver Hoef J M, Cressie N 1993 Multivariable spatial prediction. *Mathematical Geology* 25: 219–40

Wartenberg D 1985 Multivariate spatial correlation: a method for exploratory geographical analysis. *Geographical Analysis* 17: 263–83

Wegman E 1995 Huge data sets and the frontiers of computational feasibility. *Journal of Computational and Graphical Statistics* 4: 281–95

Wise S, Haining R, Ma J 1997 Regionalisation tools for the exploratory spatial analysis of health data. In Fischer M M, Getis A (eds) *Recent developments in spatial analysis*. Berlin, Springer: 83–100

Zhang A, Yu H, Huang S 1994 Bringing spatial analysis techniques closer to GIS users: a user-friendly integrated environment for statistical analysis of spatial data. In Waugh T C, Healy R G (eds) *Advances in GIS research*. London, Taylor & Francis: 297–313

Zhang Z, Griffith D 1997 Developing user-friendly spatial statistical analysis modules for GIS: an example using ArcView. *Computers, Environment and Urban Systems* 21: 5–29

6

Building Automated Geographical Analysis and Explanation Machines

Stan Openshaw

Summary

This chapter describes the development and structure of two exploratory automated geographical analysis systems. They are designed to be easy to use and to provide understandable results. There is an evaluation of their performance on synthetic data and comparisons are made with alternative cluster detection methods. The methods are briefly demonstrated and suggestions are made for their further development.

6.1 Background

There has been a vast explosion in the availability of geographically referenced data, because of developments in IT, the geographical information systems (GIS) revolution, the computerisation of administrative systems, the falling costs of data storage, and dramatic changes in the price–performance of most aspects of computing. As a result there is a vast and rapidly-growing geocyberspace of information that increasingly covers many aspects of modern life (Openshaw 1994; Goodchild, this volume). This geocyberspace constitutes the raw materials from which new knowledge, new concepts, and scientific discoveries are supposedly going to be created in the twenty-first century. If this dream is ever to be turned into reality, then we need new modelling and spatial analysis tools that can perform these functions.

In a geographical analysis context there are further problems that require attention. In particular, there are now many spatial databases available for analysis. It is no longer sensible to think in terms of years per spatial analysis task: rather, the

Geocomputation: A Primer. Edited by Paul A Longley, Sue M Brooks, Rachael McDonnell and Bill Macmillan.
© 1998 John Wiley & Sons Ltd.

focus has to be on rapid analysis ideally performed in near real-time with a capability to perform several spatial analysis tasks per day while the data are fresh and actionable. There has also been a global proliferation of GIS software that cannot provide much or any 'real' spatial analysis functionality. Useful spatial analysis tools have to be able to cope with both the special nature of spatial data and the prospective end-users who are not academics but practitioners of GIS who are not interested in research. The results have to be easily understood and self-evident so that they can be readily communicated to other non-experts. This need has been clearly expressed as follows: 'We want a push button tool of academic respectability where all the heavy stuff happens behind the scenes but the results cannot be misinterpreted'. (Adrian Mckeon, Infoshare: email: 1997). There is also a requirement for results expressed as pretty pictures rather than statistics. The problem at present is the absence of many spatial analysis methods that meet these design criteria.

An obvious solution would be the development of purely automated geographical analysis methods that involve the minimum of end-user skill while being fast, efficient, cheap and easy to apply. This chapter briefly outlines two exploratory spatial analysis methods that meet these requirements. Section 6.2 describes the geographical analysis machine. Section 6.3 outlines the further development of the technology to include a degree of geographical explanation. Section 6.4 provides a brief case study and Section 6.5 outlines how to access the software and research intended to develop it further.

6.2 Geographical Analysis Machine (GAM)

6.2.1 History

The Mark 1 Geographical Analysis Machine (Openshaw et al 1987; 1988) was an early attempt at automated exploratory spatial data analysis that was easy to understand. The GAM sought to answer a simple practical question; namely, given some point referenced data of something interesting, *where* might there be evidence of localised clustering if you do not know in advance where to look through lack of knowledge of possible causal mechanism or if prior knowledge of the data precluded testing hypotheses on the database. Even more simply put: 'here is a geographically referenced database, now tell me if there are any clusters and, if so, where are they located?' The first version (GAM/1) was developed in the mid-1980s. It was a very simple method that was very computationally intensive. The algorithm is described in Box 6.1. The term 'machine' seemed appropriate because it really needed a dedicated computer to run it on. The early runs each took over one month of CPU time on a large mainframe (an Amdahl 580). Later it was run exclusively on vector supercomputers, specifically the Cray X-MP, Y-MP and Cray 2.

The GAM method was developed to analyse child leukaemia data for northern England. It easily spotted the suspected Sellafield Cluster but it also found an even stronger major new cancer cluster in Gateshead. This is possibly the only instance

---

Box 6.1
GAM algorithm

The GAM algorithm involves the following steps:

Step 1 Read in *X,Y* data for population at risk and a variable of interest from a GIS
Step 2 Identify the rectangle containing the data, and identify starting circle radius and degree of overlap
Step 3 Generate a grid covering this rectangle so that circles of current radius overlap by the desired amount
Step 4 For each grid-intersection generate a circle of radius *r*
Step 5 Retrieve two counts for the population at risk and the variable of interest
Step 6 Apply some 'significance' test procedure
Step 7 Keep the result if significant
Step 8 Repeat Steps 5 to 7 until all circles have been processed
Step 9 Increase circle radius and return to Step 3 else go to Step 10
Step 10 Create smoothed density surface of excess incidence for the significant circles using a kernel smoothing procedure and aggregating the results for all circles
Step 11 Map this surface

Note that the original GAM/1 consisted of Steps 1 to 9; Steps 10–11 are the GAM/K version

---

of a major disease cluster being found (in this case re-discovered) by analysis (rather than journalism) since John Snow's famous cholera spatial epidemiology of the mid nineteenth century (but see Cliff and Haggett, this volume, for a wider review of epidemiological applications). Nevertheless GAM/1 was a mixed blessing! It was praised by some geographers as a major development in spatial analysis technology but it was also severely criticised by others. Additionally, the software for GAM was never distributed as (ten years ago) it was not easily run and there was therefore no purpose in disseminating it. It was, however, re-programmed in various guises and inspired a number of similar methods (Fotheringham and Zhan 1996; Wakeford et al 1996) – although some applications merely perpetuated the problems endemic to the original GAM/1 prototype and ignored subsequent improvements.

6.2.2   GAM/1: Good and Bad Aspects

GAM had a number of attractive features: in particular, it was automated, prior knowledge or ignorance was rendered equally irrelevant, it looked for localised clusters at a time when most spatial statistical methods concentrated on global measures of pattern, the search was geographically comprehensive, all locations were treated equally, spatial data imprecision was explicitly handled (a major first), the results were study region boundary invariant, the outputs were cartographic rather than expressed in terms of complex statistics, it suggested hypotheses that could be tested by other methods later, and it was an early example of a geographical data mining tool. It was suggested as a prototype of a whole class of equivalent methods since the philosophy underlying the GAM could be developed further in various ways. The principal deficiencies were: it needed a supercomputer and as a result was not easy to apply because of restricted access or long run times and there were unresolved statistical problems particularly due to multiple testing. Additionally, the tone of GAM, the high public profile of the early results, and the development of a statistical technique by a geographer upset some major statisticians who conducted a brief campaign of intensive criticism – most of which turned out to be either incorrect or irrelevant or mischievous. GAM was a deliberate attempt at automating the artistic science of statistical analysis and this was often disliked or derided as 'data trawling' because it was contrary to conventional approaches. Also the idea of localised clustering was initially but incorrectly considered to be purely a data artefact because of spatial autocorrelation. Perhaps most serious there was a failure to consider data error as a probable major cause of clustering in rare disease data.

6.2.3   Subsequent Developments

The GAM/1 was progressively developed during the late 1980s. Particularly interesting was the rotated square based method described in Openshaw et al (1989), the experiments with blob statistics described in Openshaw (1990; 1991), and the creation of 'other GAMs'; GAM/2 used circles of equal expected cases rather than distance, and GAM/3 used circles with equal observed numbers of cases. Experiments were also performed with other forms of significance testing – such as binomial, Monte Carlo and sequential Monte Carlo. However, the next major evolution of GAM into its modern form, GAM version K (GAM/K), appeared in 1990 as attempts were made to handle the problem of multiple testing in a geographical manner. There were two additional stimuli in the late 1980s: a national study of childhood cancer data involving several different research groups in the UK, each with different methods (see Draper 1991), and the IARC study of the performance of different clustering methods when applied to synthetic data (Alexander and Boyle 1996). These two projects happened in parallel and stimulated a few years of intense research, debate and mutual criticism as the various researchers using different technologies compared and contrasted their findings on both real and synthetic rare disease datasets.

6.2.4   The International Agency for Research on Cancer (IARC) Study of Clustering Methods

IARC commissioned a study in 1989–91 of all available clustering methods, many of which were associated with the early critics of GAM. Fifty synthetic cancer datasets were created for which the degree of clustering and locations of clusters were known but kept secret. These data were given to the participants who performed their analyses without any knowledge of the correct results. These methods were:

1. Potthoff–Whittingham Method (Muirhead and Ball 1989; Muirhead and Butland 1996; Potthoff and Whittingham 1996a; 1996b).
2. Cuzick–Edwards two-sample method (Cuzick and Edwards 1990; 1996a).
3. GAM-K (Openshaw and Craft 1991; Openshaw 1996).
4. Besag–Newell's Method (Besag and Newell 1991; Newell and Besag 1996).
5. ISD's Original Method (Urquhart 1988; Black et al 1991; 1996a).

This list was later extended to include four others but these were applied with knowledge of the cluster locations and of the results generated by the blind study. These were:

6. ISD revised (Black et al 1996b).
7. Cuzick–Edwards one-sample method (Cuzick and Edwards 1996b).
8. Diggle–Morris K functions (Ripley 1977; Diggle and Chetwynd 1991; Diggle and Morris 1996).
9. The CAS method (Openshaw et al 1989; Wakeford et al 1996).

The results were eventually published in 1996, although the original expected publication date was 1991: see Alexander and Boyle (1996). It was anticipated that the statistical methods preferred by the critics of GAM would work best and that this definitive study would 'kill off' the notion of geographical analysis machines forever.

6.2.5 Detection of Clustering

It is often regarded as being interesting to know whether or not particular disease data tends to show signs of clustering. Some think that a global test of clustering is, therefore, a useful piece of information. The answer is essentially either 'yes' or 'no' whether the null hypothesis of no significant clustering can be accepted or rejected. From a geographical perspective this is not particularly interesting because: (a) it is a whole map summary conclusion; (b) it says nothing about geographical location of the clusters that form the clustering; (c) many global statistics of spatial pattern are affected by the scale of the data and the choice of study region boundaries; and (d) it conveys very little useful information. Nevertheless, all the methods involved in the IARC study could yield a clustering yes/no outcome. In the GAM clustering is present if clusters exist. This is regarded as a far more sensible approach to the problem, because the nature or patterns of the distribution of clusters also tells you far more than a yes/no decision: for example, how many clusters, their spatial extent, and their pattern.

Table 6.1 summarises the results for the 10 random datasets. These data were created to represent a purely random distribution so any clusters found may be regarded as false. The best blindly applied methods have two false positives (GAM is one of them), whereas one of the later methods managed a perfect performance (Diggle–Morris). Table 6.2 shows the results for the 13 datasets that each contained

Table 6.1   *False clustering found in random data*

Method	Clusters found
Potthoff–Whittingham	2
Cuzick–Edwards	2
GAM-K	2
Besag–Newell	8
ISD original	5
ISD revised	4
Cuzick–Edwards	1
Diggle–Morris	0
CAS	4
Number of datasets	10

Note: In this table and Table 6.2 and 6.3 italicised methods were performed with knowledge of the results

Table 6.2   *Clustering found in data with one cluster in it*

Method	Clusters found
Potthoff–Whittingham	2–6
Cuzick–Edwards	7
GAM-K	11
Besag–Newell	8
ISD original	8
ISD revised	8
Cuzick–Edwards	8
Diggle–Morris	2
CAS	11

one cluster. The best two methods are GAM and a derivative (CAS), and the worst is the Diggle–Morris method. A similar pattern is evident in Table 6.3 which show the results for 2, 8 and many clusters. If the errors in these three tables are aggregated then the overall clustering test performances are those listed in Table 6.4. The superior performance of the two GAMs (GAM/K, CAS) are apparent.

6.2.6   Detection of Cluster Locations

A more important but much harder problem is to find the location of the clusters. Cancer is a rare event and the synthetic data contained considerable amounts of random noise as well as variable amounts of structure. Not all of the apparent cluster locations were findable and it is impossible to know whether the theoretical degree of clustering as input into a synthetic dataset is in fact present in the resulting synthetic data. Alexander and Boyle (1996) attempted to resolve these problems by listing all the clusters found by the three (out of the nine) methods able

Table 6.3 *Clustering found in other datasets*

Method	Number of clusters		
	2	8	Many
Potthof–Whittingham	1–8	4	6
Cuzick–Edwards	9	4	6
GAM-K	11	5	8
Besag–Newell	9	4	7
ISD original	10	5	7
ISD revised	10	5	7
Cuzick–Edwards	10	4	7
Diggle–Morris	5	4	2
CAS	12	5	8
Numbers of datasets	12	7	8

Table 6.4 *Performance of clustering test methods*

Rank of method	Percentage errors
1 GAM-K	14
2 CAS	16
3 Cuzick–Edwards one sample	24
4 ISD revised	28
5 ISD original	30
6 Cuzick–Edwards two sample	32
7 Potthoff–Whittingham	34
8 Besag–Newell	40
9 Diggle–Morris	46

Note: Based on 50 datasets

to find clusters, and then counting how many were in common. The results are shown in Table 6.5 for 1, 2 and 8 cluster datasets. The Besag–Newell method seems to do best but, as Table 6.6 shows, it also reports large numbers of false clusters, making it very unreliable. Table 6.7 reports the positive sensitivities, which is the number of times correct clusters were found after adjustment is made for the failures. The good performance of GAM/K is now evident.

6.2.7 Evaluating GAM/K as a Cluster Finder

Perhaps to the surprise of Alexander and Boyle, GAM/K was shown to be the best or equivalent best means of *both testing for the presence of clustering* and for *finding the locations of clusters*. There even appeared to be an inverse relationship between the strength of historic GAM criticism and the empirical performance of the methods preferred by the critics. The widely favoured Bayesian Mapping (viz. Kaldor and Clayton 1989) was dropped altogether as it was highly misleading. The

Table 6.5 *Finding cluster locations. Findable sites found 'near' real parent locations*

	1 parent	2	8
Besag–Newell	10	20	17
GAM/K	10	16	12
Cuzick–Edwards	7	10	16

Table 6.6 *Finding false clusters sites found 'near' parent locations*

	Random	1	2	8
Besag–Newell	8	10	5	3
GAM/K	2	3	1	0
Cuzick–Edwards	2	2	3	0
Totals	10	13	12	7

Table 6.7 *Estimated positive sensitivities*

Method	Sensitivity (%)
Besag–Newell	36
Cuzick–Edwards	66
GAM/K	87

ISD's methods were poor and unreliable clustering detectors and were not in any case designed to find clusters, just test hypotheses, while the Besag–Newell modified GAM was untrustworthy because of its high false alarm rate. Finally, Diggle's much vaunted *K* functions were next to useless for non-random data. The key point here is that rare disease data are very hard to analyse. Most of the more general spatial analysis needs in GIS will be easier. So a method that works well on rare disease data might be expected to perform even better on crime data or burst water pipes or telephone faults, etc.

Alexander and Boyle (1996: 157), the authors of the IARC study, concluded: 'The GAM has potential applications in this area if adequate computer resources are available. At the present time, however, the new, more sophisticated version of the GAM is complex, difficult to understand . . .'. That was in 1991 when there were two remaining criticisms: GAM needed a supercomputer to run it and GAM/K was complex. The question is, are these criticisms still valid today?

6.2.8 Reviving GAM/K

GAM/K still runs on the latter-day version of the Cray X-MP vector super-computer (the Cray J90). However, a single UK run was estimated to need nine days of CPU time on a single J90 processor or one hour on a 512 processor Cray T3D. This would hardly constitute a generally applicable and easy to use method relevant to GIS! It was important to make it run much faster if GAM was to have any chance of being more widely applicable. Subsequent modifications to the spatial data retrieval algorithm used in GAM/K reduced the 9 days to 714 seconds on a workstation, mainly by using extra memory to reduce the computational task. On a test problem using 1991 Census data for 150 000 census enumeration districts the total amount of arithmetic being performed was reduced from 10 498 million to 46 million floating point operations. As a result, GAM/K is now a practical tool that can be run on ordinary PCs and workstations.

6.2.9 How does GAM/K Work?

The current GAM/K consists of a second stage added to the original GAM outlined in Box 6.1. The 'significant' circles are converted into a smooth excess incidence density surface using a kernel estimation procedure (hence the K in GAM/K). An Epanechnikov (1969) kernel (see Silverman 1986) is used with a bandwidth set at the circle size and the excess incidence is smoothed out over this region. The effects are then aggregated and stored as a raster density surface. The resulting accumulation of evidence is used as the basis for conclusions about the existence, strength and locations of possible clusters. In most instances simply eye-balling these results will be sufficient to inform about the existence, strength and locations of any apparent clusters. Alternatively, a simple expert system could be used to automate the interpretation (see Openshaw and Craft 1991).

The choice of significance test is not considered as being too critical. The aim is not to test conventional hypotheses but merely determine whether or not an observed positive excess incidence is sufficiently large to be unusual and hence of interest. It is more a measure of 'unusualness' or 'surprise' than a formal statistical significance test. A number of different measures of unusualness can be applied depending on the rarity of incidence of interest; e.g. Poisson, binomial, bootstrapped z scores, and Monte Carlo tests based on rates. The aim here is not a formal test of significance, instead 'significance' is being used only as a descriptive filter employed to reject circles. It is the map created by the overall distribution of significant circles that is of most interest.

Finally, some of the critics of GAM argue that any clusters found on the output map could well be the consequence of testing multiple hypotheses. Their argument is as follows: if you set some arbitrary significance threshold (e.g. alpha = 0.05) and if you test 100 hypotheses then 5 ($100 \times 0.05 = 5$) will be false positives (i.e. they will incorrectly appear as being significant). If you test 1 000 000 hypotheses then on average 50 000 will appear as false positives. Two considerations nevertheless weaken this argument: it assumes the hypotheses are independent whereas in a GAM search they are clearly not (because the circles overlap) and it ignores the geography of the problem as it is surely quite different if all the significant circles occur around one or two locations rather than be scattered randomly all over the map. These effects can be studied by Monte Carlo simulation and this feature is now built into GAM.

6.3   Building a Geographical Explanations Machine

6.3.1   Background

GAM is purely a pattern detector and no explanation is provided that can help the user understand what variables may be geographically correlated with any clusters it may find. It is also possible that the apparent patterns being uncovered by GAM merely reflect missing confounding variables and that they are, therefore, of no real consequence. However, there are many different possible confounding variables that may be considered important. Some of these could well be related primarily to individual behaviour, while others may principally be artefacts of geographical location; some may be both. This is particularly the case in studies of disease, where some important variables relate to the unmeasured individual exposure to allegedly harmful chemicals. Yet many of these variables 'thought to be useful' in developing a better understanding of a disease may well be missing from spatial databases and they would have to be specially collected or measured. However, they may still be implicitly represented by map location. Indeed it was precisely to try to exploit the latent potential of the map (and the associated GIS databases) as a source of 'geographical explanation' that the geographical correlates exploration machine (GCEM) of Openshaw et al (1990) was developed. The hope is that geographical information may provide useful surrogate information to compensate for the lack of better and more relevant data and thus help guide any subsequent more detailed inquiries. GCEM attempts to automate the step after GAM whereby the investigator imports the GAM clusters into a GIS and starts to look for possible linkages with various data layers via multiple spatial queries. This is one of the oldest forms of geographical analysis and it predated GIS by about a century: see, for example, the early geographical disease atlas of Haviland (1892). However, what GIS has so far failed to do is to extend or improve the sophistication of the process of searching for potential geographical associations.

6.3.2   A Geographical Correlates Exploration Machine

The geographical correlates exploration machine (GCEM) of Openshaw et al (1990) attempts to automate the search for geographical correlates (i.e. localised spatial associations) between clusters and geographical variables that was suitable for use within GIS environments. The problem of knowing which of M^{K-1} possible permutations of M coverages with K overlay operations will yield the 'most interesting' map patterns is solved by enumerating and evaluating as large a subset as possible. The hope is that this brute force computational approach will compensate for the user's lack of knowledge to achieve a degree of artificial intelligence (Openshaw and Openshaw 1997). The original GCEM algorithm is described in

---

Box 6.2
The GCEM algorithm

Step 1 Consider a permutation of any of the M coverages
Step 2 Overlay these coverages to yield a new composite coverage, i.e. a polygonal map that defines zones
Step 3 Use a point in polygon method to assign the data to the zones formed in Step 2 to create an aggregate dataset
Step 4 Test the results for each of the zones and flag those that have unusually high incidence levels together with details of the coverages and categories that created and characterise it
Step 5 Repeat Steps 1 to 4 for all 2^{M-1} permutations, although in practice computer time restrictions sets upper limits on the maximum numbers of coverages that can be overlaid because of a combinatorial explosion once M and or K become 'large'
Step 6 Rank the 'significance' of the polygon fragments. Examine the results and the associated coverage details for evidence of recurrence, geographical concentration insight and interest. Note that there is sufficient information available here for any polygon of interest that creates an unusual result to be readily re-created via a series of GIS queries, so that a GIS might be extremely useful at this point in the search as a means of adding detail and testing out ideas. In this type of exploratory study the user is really functioning as a map detective on the trail of some potentially interesting map patterns using spatial information as clues to aid the investigation

---

Box 6.2. Note that GCEM is a highly computationally intensive process. For example, if there are 15 coverages (M) and all permutations with up to five different overlays (K) are to be considered, then there are 4943 different sets of polygons to be examined, each requiring a polygon overlay operation and an assignment of N data points via a point in polygon procedure.

The main problems with GCEM in the late 1980s were: it needed a supercomputer to power it; it was better suited to a parallel supercomputer, but only vector machines were available; it suffered from the same range of statistical and other problems as GAM, particularly multiple testing; and the virtual map overlay simulator used in Openshaw et al (1990) does not yield accurate results for non-convex polygons (although this is not too serious given the nature of the search). Practical experience with GCEM was also very limited. Additionally, there was no explicit control over the nature of the areas being examined for patterns as these were defined by the overlaying of polygonal maps. Some would argue that GCEM was critically flawed in attempting to find patterns that could be explained in a 'geographical way' – viz. spatially concentrated and locally associated with a particular combination of map features. Some would consider this to be a major restriction yet it is an endemic and unavoidable characteristic of geographical inquiry. In essence the map is being used as a surrogate for other missing information, although it is not being used as an explanatory variable in its own right (that would just be too naive for words). Maps do not cause patterns to appear by themselves but they almost certainly contain strong clues to other variables that do, if only we were clever enough to decode them! It is simultaneously a great strength and a weakness. You also have to be realistic!

6.3.3   Building a Geographical Explanations Machine

The underlying idea of having a GCEM is still a good one, but it just needs re-developing into a more widely useable form and fresh thought should be given to what functions it should seek to perform. An obvious development is to link GAM with GCEM. Instead of overlaying complete polygon coverages the GCEM method can be re-cast in terms of the moving circle window used in GAM. The GAM circle defines a set of arbitrary local search regions within which the equivalents of 2^{M-1} coverage permutations can be examined as local covariates. This approach helps unify the conventional epidemiological approach with GAM but does so in the GIS environments previously inhabited by GCEM. There is no suggestion that patterns are 'caused' by what may be termed 'GIS variables', only that if 'strong' or 'significant' or localised or 'recurrent' geographical associations can be found then this may well be of interest to cluster or pattern hunters as an indicator of where to look further. This GAM–GCEM hybrid is renamed a geographical explanations machine (GEM). It serves two seemingly contradictory objectives:

1. To find significant localised excesses that can only be ascribed to what might be regarded as mystery locational factor X because they cannot be made to go away no matter how hard the attempt using covariates based on a wide range of map (and any other) spatial datasets.
2. To find significant localised excesses that are (perhaps locally) spatially associated with particular sets or combinations of geographical variables in a sufficiently strong way to offer a form of geographical explanation of them and hence appear as worthy of further study.

The secret of GEM is how to build in these geographical covariates. This is straightforward but it can be confusing. The idea is to adjust expected rates for the incidence of a variable within a GAM circle to allow for local geographical covariation. In a GIS there are often additional levels of contextual information that may provide useful surrogates for missing variables. For example, points can be classified by their location with respect to various data layers; viz. geological type and membership of a point- or line-based buffer region around a chemical works or a road. This standardisation for geographical covariates can be done as follows. Consider a single coverage C which has five categories. The adjusted incidence rate is

$$\sum_{j=1}^{5} P_j \, C_j \qquad (1)$$

where P_j is the population at risk inside the circle in the jth category of the coverage and C_j is the average rate for this category in the coverage. This readily generalises to two-way interactions by adding subscripts so if a second subscript k is used to represent the effects of a second coverage, this time with eight categories, then equation (1) becomes

$$\sum_{j=1}^{5}\sum_{k=1}^{8} P_{jk}\, C_{jk} \tag{2}$$

where P_{jk} is now the population at risk inside the circle in the jth category for coverage 1 and in the kth category on coverage 2, and C_{jk} is the associated incidence rate. Note that the composition of C_{jk} and P_{jk} now depends on which of the two from M coverage permutations are being considered. For example if there are ten coverages then there are 45 unique permutations of two-way interaction effects to be considered. This approach can be extended to consider three and four or higher order interaction effects; however, there may be little merit in going beyond three-way interactions unless the database is extremely large because of small number effects. Computation times also explode exponentially so this GEM is living on the edge of a combinatorial precipice. Table 6.8 gives some relevant counts. Note that for each of the circles being examined the number of coverage permutations given in this table has to be considered. The use of supercomputers merely gains you a few hundred thousand more permutations before computation times will again become impossibly large. So this technology has hard limits to the degree of searching it can do. However, it is thought that these should not present any significant practical constraints because the principle of parsimony and data sparsity sets in well before the combinatorial explosion become critical. The counts in Table 6.8 are theoretical maximum limits and in reality only a small fraction would be relevant (because there are data) to any particular search circle. Parsimony also suggests that the number of overlaid maps K is best kept small, in which case large numbers of coverages can be examined. For example, there are only 3160 permutations of 80 maps into two overlays; compared with 24 040 016 if five overlays are considered. The latter is feasible but only on fast parallel machines, while the former is possible even on a PC workstation.

6.3.4   Four Different Search Modes

There are four ways of running GEM:

Mode 1:   *Pure GAM.* This is a GAM run with no coverage information being used and will only detect patterns. If some patterns are found then modes 2, 3 and 4 are worth considering.

Mode 2:   *GAM with geographical covariates.* The aim now is to include 1, 2, 3, . . . way interaction effects with M available geographical covariates. When a circle is found which GAM would regard as having a significant excess, then the combinations of coverages are examined to try to remove the significance and thus 'explain away' the clustering. Surviving circles should be examined further for possible related variables, or at the least be mapped, because they cannot so easily be made to go away. However, there may also be a case for looking at the circles that have been explained away to try to establish the reasons and the coverages responsible.

Table 6.8 Number of coverage permutations

Total number of coverages, M	Level of interaction, K							
	1	2	3	4	5	6	7	8
1	1	–	–	–	–	–	–	–
5	5	10	10	5	1	0	–	–
10	10	45	120	210	252	210	120	45
15	15	105	455	1365	3003	5005	6435	6435
20	20	190	1140	4845	15504	38760	77520	125970
25	25	300	2300	12650	53130	177100	480700	1081575
30	30	435	4060	27405	142506	593775	*	*
40	40	780	9880	91390	658008	*	*	*
50	50	1225	19600	230300	2118760	*	*	*
60	60	1770	34220	487635	5461512	*	*	*
70	70	2415	54740	916895	12103014	*	*	*
80	80	3160	82160	1581580	24040016	*	*	*
90	90	4005	117480	2555190	43949268	*	*	*
100	100	4950	161700	3921225	75287520	*	*	*
150	150	11750	551300	*	*	*	*	*
200	200	19900	1313400	*	*	*	*	*

Mode 3: GAM as a GCEM. Another way of running GAM is to change the Mode 2 operation so that it replicates the intention inherent in the original GCEM. The original GCEM examined polygons formed by permutations of overlays for excess incidence. These polygons are homogenous with respect to the GIS overlays that defined them so note could be made of the GIS data most associated with an excess value. Here this process is modified so that the area being examined is defined by a circle and its interaction with the implicit but invisible underlying map polygons. Not all the polygons may lie completely within the circle or the circle may contain fragments of several different ones. It is necessary, therefore, to classify the cases inside each circle being examined into homogenous subsets (so their GIS features can be unambiguously defined), and then perform analysis on all subsets large enough to be considered safe to use. All circles with map permutations that yield significant results are now output together with their associated map features. This replicates the type of spatial query that GIS users would do when presented with evidence of a cluster, except that the computer search for potentially interesting results is automated and far more comprehensive than could be performed manually.

Mode 4: Mixed search. A further possibility is to combine Modes 2 and 3. Mode 2 identifies 'clusters' that are covariate proof in that they cannot be made to go away, but says nothing about their geographical characteristics other than location. Of particular interest is the prospect that many or some of these 'clusters' may have common features that if known to the user may trigger off further investigations. The Mode 2 approach is used to identify the least significant results for the homogenous data subsets used in Mode 3. This essentially identifies GAM-significant circles where the addition of geographical covariates is used to try to make the clusters defined by GCEM (Mode 3) go away.

6.4 Case Study

Both GAM and GEM are now available on the World Wide Web. The real computing 'machine' has been replaced by an Internet URL! You can now send your data to Leeds and obtain results back for local analysis. To illustrate the self-evident nature of the GAM and GEM, consider the following very brief exploratory spatial analysis of long-term limiting illness data for northeast England: see Openshaw et al (1998) for details. The 1991 Census data for long-term limiting illness is analysed for 6905 Census Enumeration Districts. These Census data are considered to be far easier to analyse than rare disease data and far safer in that the data are not highly confidential in nature. However, long-term limiting illness is not a rare disease, so a Monte Carlo significance test procedure is used and age–sex covariates used to modify the expected rates. The GAM/K (or GEM mode 1) results in Figure 6.1 show where there are localised excesses. The more intense the excess the stronger the clustering. The results are self-evident. However, if necessary

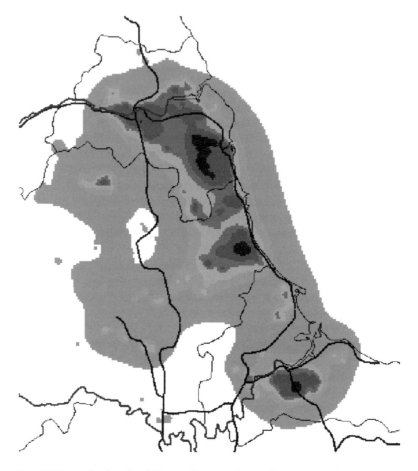

Figure 6.1   GAM results: localised illness clustering for northeast England

then Monte Carlo simulation of multiple testing can be performed and here it is readily apparent that these results are not attributable to this factor.

Six pseudo GIS coverages are created that reflect the following predictor variables that may be associated with long-term limiting illness:

- population density of containing Ward
- deprivation index of containing Ward
- unemployment in Census Enumeration District
- crowding in Census Enumeration District
- single parents in Census Enumeration District
- social class I in Census Enumeration District

These are pseudo coverages (rather than the more usual map layers) to illustrate how it is possible to build relationships into GEM. An approach is to compute 5 km counts around each Census Enumeration District and recode them as

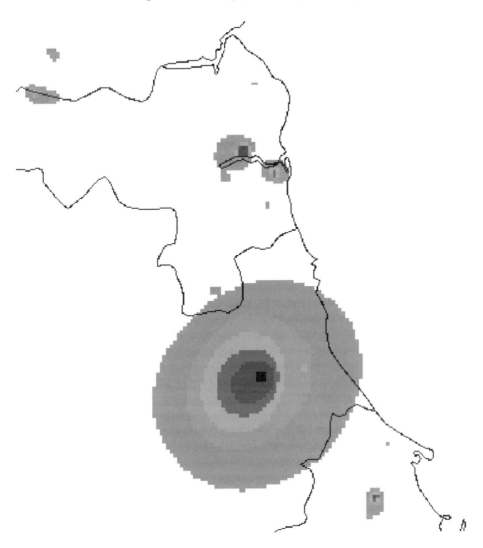

Figure 6.2 *GEM mode 2 results: clustering that is unexplained by geographical covariates*

quintiles (thereby effectively creating a pseudo coverage with five categorical values). This is similar to using a choropleth map and then assigning each Census Enumeration District to the relevant class value. Here 187 permutations of the six GIS pseudo coverages examined for each of many million circles takes a few days on a slow workstation. Figure 6.2 shows the GEM mode 2 results that attempt to destroy the clustering. It succeeds but leave one major cluster unexplained. In Figure 6.3, there is a mode 3 run showing where the clusters can be most readily explained away. Subsequent examination of which pseudo-coverage combinations match the clustering maybe helpful as a means of understanding the map patterns.

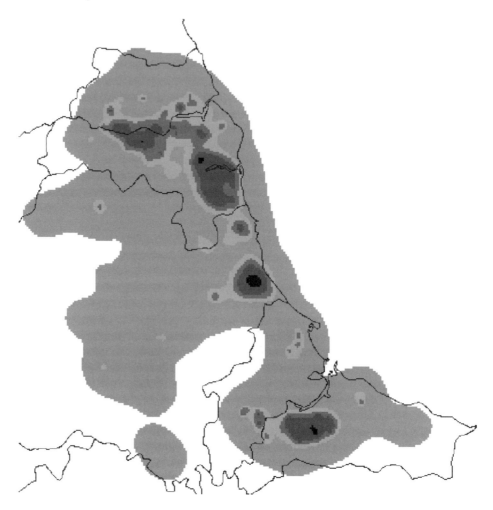

Figure 6.3   *GEM mode 3 results: clustering that can be explained by geographical covariates*

6.5   Conclusions

A GEM has been built as a GAM–GCEM hybrid. It might well be the first real
geographical analysis and explanation machine to be developed. It explores point
data for evidence of localised patterns and then attempts to 'explain' the excesses in
a geographical way. It is generically applicable since its function is data independent. It can be run on PCs, UNIX workstations, and if required on parallel
supercomputers. The next step in the longer term evolution of GEM is to change
the nature of the brute force dumb map searches used here so that equivalent
results can be produced more speedily using smart search methods. It is also useful
to consider how to create opportunities for algorithmic visualisation, and also to
expand the number of database dimensions that can be simultaneously explored.

Meanwhile here is a new and widely applicable geographical analysis tool that can be applied to an increasing range of GIS data. Both GAM/K and GEM are useful map explorers in their own right as well as providing a useful test-bed for subsequent developments based on more intelligent search technologies – as, for example, with MAPEX (Openshaw and Perrie 1996) or the Space Time Attribute Creature (STAC) of Openshaw (1994b; 1995). Both GAM and GEM are available for use over the Internet.

References

Alexander F E, Boyle P 1996 *Methods for investigating localised clustering of disease*. Lyon, IARC Scientific Publications No. 135

Besag J, Newell J 1991 The detection of clusters in rare disease. *Journal of the Royal Statistical Society A* 154: 143–55

Black R J, Sharp L, Urquhart J D 1991 An analysis of the geographical distribution of childhood leukaemia and non-Hodgkin lymphomas in Great Britain using areas of approximately equal population size. In Draper G J (ed.) *The geographical epidemiology of childhood leukaemia and non-Hodgkin lymphoma in Great Britain 1966–1983*. London, HMSO: 61–8

Black R J, Urquhart J D, Kendrick S W, Bunch K J, Warner J, Jones D A 1992 Incidence of leukaemia and other cancers in birth and schools cohorts in the Dounreay area. *British Medical Journal* 304: 1401–5

Black R J, Sharp L, Harkness E F, McKinney P A 1994 Leukaemia and non-Hodgkin's lymphoma – incidence in children and young adults resident in the Dounreay area of Caithness Scotland 1968–91. *Journal of Epidemiology and Community Health* 48: 232–6

Black R J, Sharp L, Urquhart J D 1996a Analysing the spatial distribution of disease using a method of constructing geographical areas of approximately equal population size. In Alexander F E, Boyle P (eds) *Methods for investigating localised clustering of disease*. Lyon, IARC Scientific Publications No. 135: 28–39

Black R J, Sharp L, Urquhart J D 1996b Extension of the ISD method. In Alexander F E, Boyle P (eds) *Methods for investigating localised clustering of disease*. Lyon, IARC Scientific Publications No. 135: 186–99

Cuzick J, Edwards R 1990 Spatial clustering for inhomogeneous populations with discussion. *Journal of the Royal Statistical Society B* 52: 73–104

Cuzick J, Edwards R 1996a Clustering based on K nearest neighbour distributions. In Alexander F E, Boyle P (eds) *Methods for investigating localised clustering of disease*. Lyon, IARC Scientific Publications No. 135: 55–67

Cuzick J, Edwards R 1996b Cuzick-Edwards one-sample and inverse two-sampling statistics. In Alexander F E, Boyle P (eds) *Methods for investigating localised clustering of disease*. Lyon, IARC Scientific Publications No. 135: 200–2

Diggle P J, Chetwynd A G 1991 Second-order analysis of spatial clustering for inhomogeneous populations. *Biometrics* 47: 1155–63

Diggle P, Morris S 1996 Second order analysis of spatial clustering. In Alexander F E, Boyle P (eds) *Methods for investigating localised clustering of disease*. Lyon, IARC Scientific Publications No. 135: 207–14

Draper G (ed) 1991 *The geographical epidemiology of childhood leukaemia and non-Hodgkin lymphomas in Great Britain 1966–83*. Studies in Medial and Population Subjects No. 53, OPCS, London, HMSO

Epanechnikov V A 1969 Nonparametric estimation of a multidimensional probability density. *Journal of Theoretical Probability Applications* 14: 153–8

Fotheringham A S, Zhan F B 1996 A comparison of three exploratory methods for cluster detection in spatial point patterns. *Geographical Analysis* 28: 200–18

Haviland A 1892 *The geographical distribution of disease in Great Britain.* London, Swan Sonnenschein

Kaldor J, Clayton D 1989 Role of advanced statistical techniques in cancer mapping. In Boyle P, Muir C S, Grundmann E (eds) *Cancer mapping.* Heidelberg, Springer: 87–99

Lolonis P 1995 The exploratory spatial analysis of birth defect rates in an urban population. *Statistics in Medicine* 15: 717–26

Muirhead C R, Ball A M 1989 Contribution to the discussion of the Royal Statistical Society meeting on Cancer near Nuclear Installations. *Journal of the Royal Statistical Society A* 152: 376

Muirhead C R, Butland B K 1996 Testing for over-dispersion using an adapted form of the Potthof–Whittinghill method. In Alexander F E, Boyle P (eds) *Methods for investigating localised clustering of disease.* Lyon, IARC Scientific Publications No. 135: 40–52

Newell J N, Besag J E 1996 The detection of small area database anomalies. In Alexander F E, Boyle P (eds) *Methods for investigating localised clustering of disease.* Lyon, IARC Scientific Publications No. 135: 88–100

Openshaw S 1990 Automating the search for cancer clusters: a review of problems, progress and opportunities. In Thomas R W (ed) *Spatial epidemiology.* London, Pion: 48–78

Openshaw S 1991 A new approach to the detection and validation of cancer clusters: a review of opportunities, progress and problems. In Dunstan F, Pickles J (eds) *Statistics in Medicine.* Oxford, Clarendon Press.

Openshaw S 1994a Computational human geography: exploring the geocyberspace. *Leeds Review* 37: 201–20

Openshaw S (1994b) Two exploratory space–time attribute pattern analysers relevant to GIS. In Fotheringham S, Rogerson P (eds) *GIS and Spatial Analysis.* London, Taylor & Francis: 83–104

Openshaw S 1995 Developing automated and smart spatial pattern exploration tools for geographical information systems applications. *The Statistician* 44: 3–16

Openshaw S 1996 Using a geographical analysis machine to detect the presence of spatial clusters and the location of clusters in synthetic data. In Alexander F E, Boyle P (eds) *Methods for investigating localised clustering of disease.* Lyon, IARC Scientific Publications No. 135: 68–87

Openshaw S, Craft A 1991 Using the Geographical Analysis Machine to search for evidence of clusters and clustering in childhood leukaemia and non-Hodgkin lymphomas in Britain. In Draper G (ed.) *The geographical epidemiology of childhood leukaemia and non-Hodgkin lymphoma in Great Britain 1966–1983.* London, HMSO: 109–22

Openshaw S, Openshaw C 1997 *Artificial intelligence in geography.* Chichester: John Wiley

Openshaw S, Perrie T 1996 User centred intelligent spatial analysis of point data. In Parker D (ed) *Innovations in GIS 3.* London, Taylor & Francis: 119–34

Openshaw S, Charlton M, Wymer C, Craft A W 1987 A Mark I Geographical Analysis Machine for the automated analysis of point datasets *International Journal of Geographical Information Systems* 1: 335–58

Openshaw S, Charlton M, Craft A W, Birth J M 1988 Investigation of leukaemia clusters by the use of a Geographical Analysis Machine. *Lancet* I: 272–3

Openshaw S, Wilkie D, Binks K, Wakeford R, Gerrard M H, Croasdale M R 1989 A method of detecting spatial clustering of disease. In Crosbie W A, Gittus J H (eds) *Medical response to effects of ionising radiation.* London, Elsevier: 295–308

Openshaw S, Cross A, Charlton M 1990 Building a prototype Geographical Correlates Exploration Machine. *International Journal of Geographical Information Systems* 3: 297–312

Openshaw S, Turton I, Macgill J 1998 Using the Geographical Analysis Machine to analyse census long term limiting illness data. *Geographical Systems* (in press)

Potthoff R F, Whittinghill M 1966a Testing for homogeneity I. The binomial and multinomial distributions. *Biometrika* 53: 167–82

Potthoff R F, Whittinghill M 1966b Testing for homogeneity II. The Poisson Distribution. *Biometrika* 53: 183–90

Ripley B D 1977 Modelling spatial patterns (with Discussion). *Journal of the Royal Statistical Society B* 39: 172–212

Silverman B W 1986 *Density estimation for statistics and data analysis.* London, Chapman & Hall

Urquhart J 1988 Exploring small area methods. In Elliott P (ed.) *Methodology of enquiries into disease clustering.* London, Small Area Health Statistics Unit, London School of Hygiene and Tropical Medicine: 41–51

Wakeford R, Binks K, Gerrard M, Wood A 1996 The CAS method. In Alexander F E, Boyle P (eds) *Methods for investigating localised clustering of disease.* Lyon, IARC Scientific Publications No. 135: 219–26

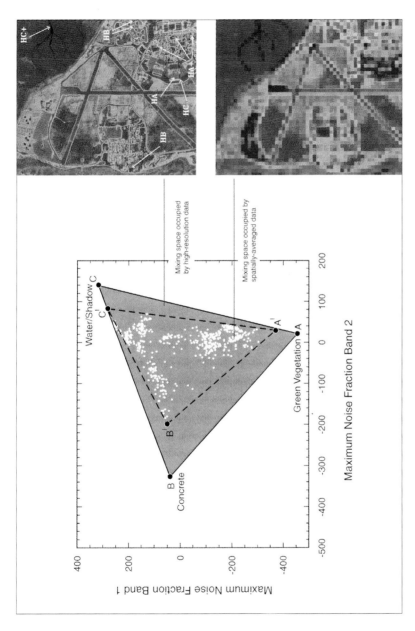

Plate 1 An example of pixel unmixing using CASI data from southern England. The ten CASI bands have been reduced to the first two maximum noise fraction (MNF) bands and the mixing spaces occupied by image data with two spatial resolutions are shown. The triangular mixing space ABC encompasses the high spatial resolution data (2 m pixel size), while data from the coarser spatial resolution image (36 m pixel size) all fall within mixing space A'B'C'. Endmembers A, B and C could be used to estimate the proportions of green vegetation, concrete and water/shadow in each of the pixels in the coarse spatial resolution image

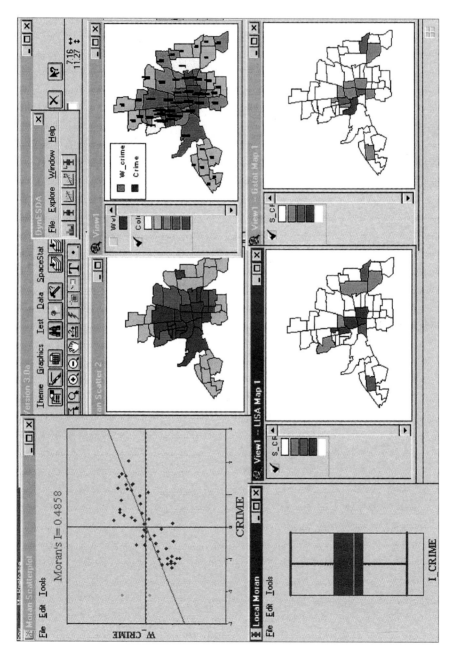

Plate II *Dynamically linked windows and visualisation of spatial autocorrelation in the SpaceStat extension for ArcView. Details are discussed in the text*

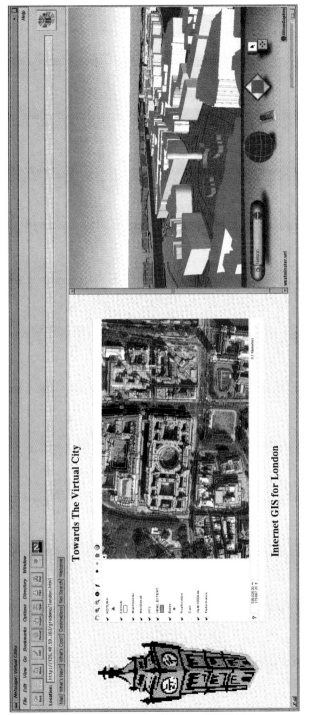

Plate III *The Internet GIS for London in the VR Theatre full-screen mode*

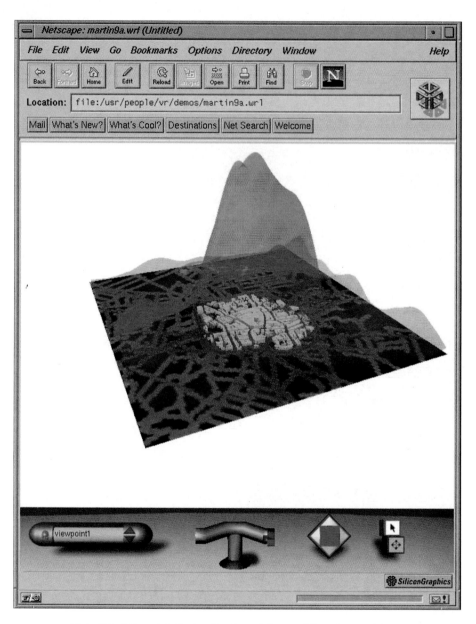

Plate IV *2-D and 3-D visualisation of the retail location problem*

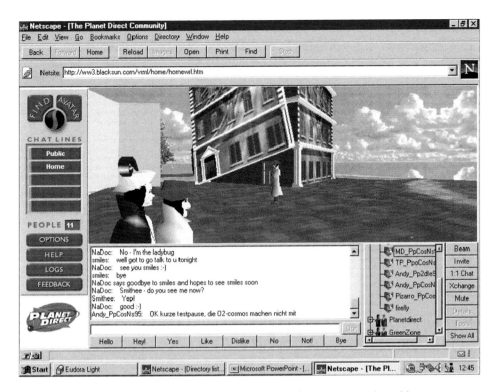

Plate V   Avatars manipulating 3-D built form in a virtual world

Plate VI   Manipulating a 'bouncing ball' in a photorealistic panorama

Plate VII Virtual urban design in Wired Whitehall

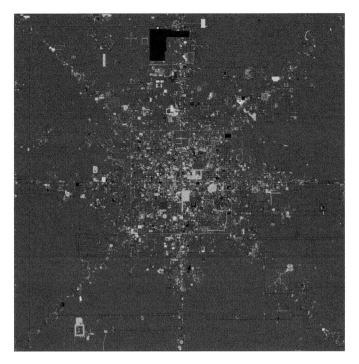

Plate VIII   *The morphology of AlphaWorld*

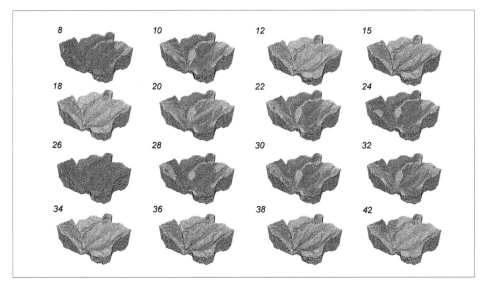

Plate IX   *Sequential plots of surface discharge for the Catsop runoff scenario. Scale indicates relative amounts of surface run-off, blues – much runoff, yellows – little or no runoff*

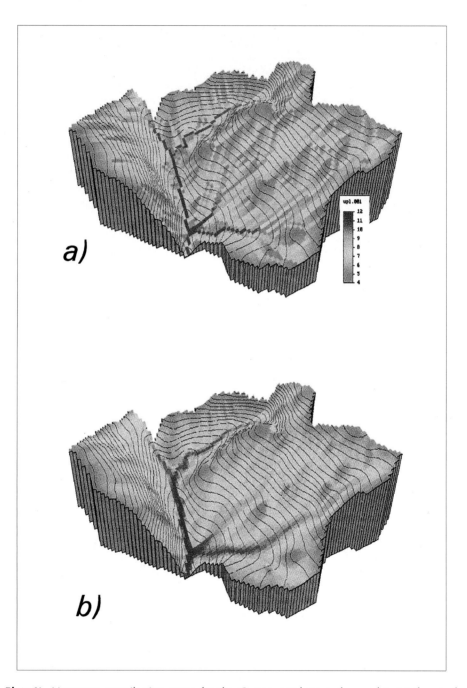

Plate X *Upstream contributing areas for the Catsop catchment (log scale): (a) the single solution for the thin plate spline surface; (b) mean result obtained from 100 simulations with an RMS error of ± 10 cm added to the DEM*

Part Four
REAL AND VIRTUAL ENVIRONMENTS

7

Visualising Different Geofutures

Keith C Clarke

Summary

Technological development often precedes science, and a core objective of this chapter is to illustrate the ways in which new computer hardware and truly portable computing will lead the future development of geocomputation. A second theme is the new means of human–computer interaction that these systems require and the ways in which the technology of wearable and other field computing will usher in a new era of field exploration in geography. This chapter also examines the prototypes of these systems and investigates the range of sensory geofutures that they might generate. It assesses the demands that new hardware devices will place on software, the new applications that will become possible, and the new challenges for geocomputational research in the coming hyperinteractive information age.

7.1 Introduction

Imagine for a while the following scenario. The Santa Barbara police have received a report that a small plane that took off from the Santa Barbara airport to fly to Las Vegas has disappeared from radar in the Santa Ynez Mountains. With few roads and about 1800 m of relief, rescue access is extremely difficult. A mobile unit is dispatched with the newly available GeoSearch system. Using the in-vehicle global positioning system (GPS) receiver, the crew in the jeep on the way to the site download a real-time satellite image of the likely search area, selected on the fly using pre-stored digital topography, vegetation maps, and the road network plus the plane's radar trajectory downloaded from the airport radar. The GeoSearch operator downloads the imagery into her belt-worn mobile computer, and checks

Geocomputation: A Primer. Edited by Paul A Longley, Sue M Brooks, Rachael McDonnell and Bill Macmillan.
© 1998 John Wiley & Sons Ltd.

that her vision-controlled peripheral vision image and map display is fully functional. All looks OK as the operator stops the jeep at the point computed by a least cost path algorithm to offer the best stretcher-carry out of the densely vegetated and steep canyon where the plane is clearly visible on the satellite imagery. From the 1 cm resolution imagery, one injured person can be seen, and appears to have broken legs.

The GeoSearch operator sees the reassuring flashing orange dot of her own GPS location closing in on the image view of the location. A solid red line rises into view in the peripheral eye display indicating a dangerously steep slope to the north to be avoided as she and her search team crash through the trees. At one time in the arduous trip, the head-mounted visicam detects the characteristic leaf shape of poison oak, and warns the crew. At last the site is reached, and the second victim located. Unable to see the extent of the victim's injury, the GeoSearch operator calls up the first available trauma surgeon on her head-mounted Web-cam. He is in the Cottage Hospital cafeteria in Santa Barbara, but knows the signs of a compound fracture when he sees one. A quick diagnosis and the pertinent Red Cross emergency manual page appears in the full vision display of the operator. Twenty minutes later, at a spot recomputed on the fly as having the best combination of stretcher access and flat treeless terrain, a Medivac helicopter loads the injured and the team heads back to their jeep. The parting shot of the successful rescue team runs on the 11 o'clock news that evening, grainy through its having been shot with the infrared night camera necessary for such a night-time rescue operation.

Nothing in this scenario is futuristic. What is missing is the integration of hardware, software and data that make this geocomputational degree of interoperability a reality. From the hardware point of view, the scenario used in-vehicle navigation systems, mobile cellular or broadcast network communications based on the Internet's protocols, and some quite simple wearable computing technology. From a software point of view, the analyses performed are almost routine for geographical information and computer vision systems. From the data perspective, we assumed Web access to public domain digital cartographic data for the GIS, and also access to high-resolution satellite imagery, thermal sensing and radar data. All the scenario seems to lack in terms of credibility is the integration of these rather disparate systems, and the communications bandwidth to do the integration and communications in real time. And such systems integration seems only a few years away (Krygiel 1997).

Paul Forman (1987) has made a convincing case from the point of view of the historian of science, that in the post World War II era, technology often precedes science in a way that was unprecedented in the traditional view of scientific research. Forman's analysis examines the full social context of scientific development, including the ever-present military force driving civilian technologies, and his prevailing message has been that, in the recent past, 'invention is the mother of necessity' and 'if we build it, they will come'. Geocomputation saw such a 'hardware first' period in its early days (Coppock and Rhind 1991; Foresman 1998). A particularly good example is the early map data capture problem (Auerbach 1972; Clarke 1995), where manual digitising tablets mimicked the drafting table of the replication stage from prior technologies, and led a vector era in GIS: this was followed by the innovation of scanning technology, which ushered in the raster era.

This chapter is principally about hardware. This is because the hardware of the scenario outlined above either does not as yet exist, or it exists as a set of disjoint technologies aimed at non-spatial markets. We do not know, but perhaps the scenario is already being played out every day in the war games of the United States military, and somewhere in the back of a general's mind is a plan for 'dual-use' of the technology. Here, we will make an examination of the hardware that could make the scenario a reality. An important aspect of the present geocomputing era is geographical scale (see Cliff and Haggett, this volume). Negroponte has stated that '[t]he digital planet will look and feel like the head of a pin' (Negroponte 1995: 6), and it is this smallness that will cast the scientist back out into the field, for a second age of geographical discovery based on in-the-field application of GIS principles and methods.

Two elements permeate the hardware discussions. The first is that of ubiquitous computing. Ubiquitous computing as a phase began with the lap-top computer and the personal digital assistant, and is now approaching the reality of independent portable and fully mobile field computing. Contributing technology has included miniaturisation in electronics, nanotechnology, lasers, the global positioning system, digital cameras, and the development of interactive input and output devices from flight simulation, gaming and space science. Contributing intellectual domains have included GIS, engineering, image processing, remote sensing, ergonomics, design, bionics, computer science, scientific visualisation, and communications. Allied to this, the fledgling cyberspace bionics industry has been helped by the World Wide Web, the Internet, high-quality consumer electronics (e.g. the sound editing and recording industries), cellular communications, inexpensive GPS differential correction and carrier phase processing, and some new developments in the creation and use of innovative display methods. Side benefits have come from GIS data exchange and metadata standards, from standardisation and 'clone' sales in the computing industry, from innovations in recording and broadcast such as HDTV, and from the declassification of previously classified government technology.

In addition to the extreme portability of ubiquitous computing, a second theme is the new means of human–computer interaction these systems require. Once again, technology has led in producing hyperinteractive systems. Hyperinteractivity is a characteristic of highly mobile and multisensory input and output systems for interacting with computers and GIS. Hyperinteractive components of computing systems allow many more options for input and output with the computer than was possible with the simple (monitor/keyboard) or even the desktop (monitor/keyboard/mouse) user interface. Hyperinteractive devices are those which allow the conversion of human physical movements into digital input streams. Input can be established from voice, hand movement, head and face movement, eye movement, body movement, and more recently thought control. Output can also be multisensory, although predominantly the display component has been limited primarily to vision simulation and vision enhancement, and sound.

Together, these trends have brought forth the technology of wearable computing, and of field computing so mobile that a new era of field exploration is now possible in geography. In the following sections, we will examine the prototypes of these systems in some detail. Assuming these prototypes become more widely available, what sort of visualisations of the geofuture might they generate? The second half of this chapter extends Forman's technology hypothesis to ask not only what demands this new suite of hardware devices will place on software, but also what new applications will be possible, and what will be the new challenges for geocomputational research that face the ubiquitous and hyperinteractive geographers raised for the information age. There has been much discussion recently of augmented reality and augmented memory, of the seamless integration of simulations and projections into everyday decision making and living. Negroponte has stated that 'Computing is not about computers any more. It is about living.' (Negroponte 1995: 6). While some of the prospects are rather frightening in scope, the potential benefits are there to be seized by those willing to take up the challenges.

7.2   Future Hardware and the Hardware Future

Devices for geocomputation have so far been built around the MIMO model presented first by Tobler (1959). That is, a geographical information processing system must be capable of 'getting the Map In', processing it, and then 'getting the Map Out'. In this simple model, a core computing system based on the presence of a central processing unit with ancillary random access memory and permanent storage, is fed information from a set of input devices (Figure 7.1: see also Clarke 1995). Systems software running on the computer interacts with the user through applications software, and through hardware devices. Devices are specialised, in that some perform input and some output. In the figure, two additional types of device are necessary for spatial interaction, and the input and output demands of accuracy, precision and size associated with maps. These graphical input and output devices are controlled by the applications software and device drivers. The graphical user interface (GUI) controlling the hardware is based on the windows, icons, menus and pointers (WIMP) model. This GUI methodology has been closely allied with the desktop metaphor, in spite of several efforts to break away from it in the case of spatial data (Clarke 1997). An addition to the system, the wide area network and the Internet, was seen as an alternative method for data provision, and assists effective search (see Goodchild, this volume).

Hyperinteractivity and ubiquitous computing have identical demands to the basic model. Consider the typical wearable computing system, such as the University of Oregon's vest-worn model shown in Figure 7.2 (right). There are many different components of the MIMO model. First, the CPU (a 70 MHz Pentium processor with 40 MB RAM) is enclosed in a small box contained within a pocket in the vest, or sometimes attached to a belt. In short, the 'Computer' box in Figure 7.1 has shrunk to a space the size of a paperback book, weighing about 1 kg, the bulk of which is ports to support the various input/output (I/O) devices and batteries. A one gigabyte removable hard drive and a PCMCIA card constitute the permanent

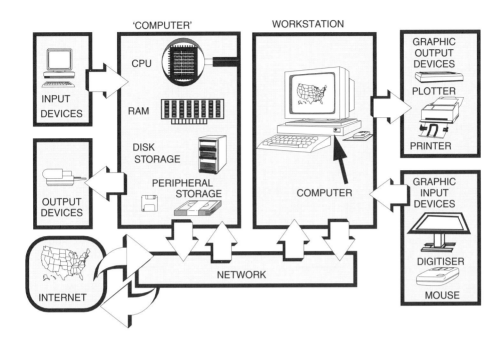

Figure 7.1   *A map of the computer. Source: Clarke (1995). Used with permission*

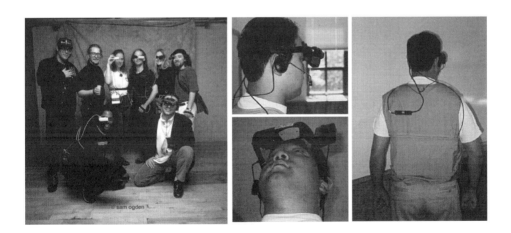

Figure 7.2   *Examples of wearable computers. Left: group picture of the MIT Wearable Computing Group. See* http://lcs.www.media.mit.edu/projects/wearables/. *Photograph copyright Sam Ogden: used with permission. Right: the University of Oregon Wearable Computer. Source:* http://www.cs.uoregon.edu/research/wearables/Oregon

memory. The other ports are two serial, two audio (in and out) and a SuperVGA video port. The network connection, backbone and Internet connection shown in Figure 7.1 is a Metricom wireless modem (TCP/IP dial up connection, 14.4 kbps), a link to the Internet which is thin, but nevertheless versatile when connected to a cellular telephone. Of particular interest, of course, are the I/O devices themselves. Output consists of a single-eye head-mounted active matrix display with 640 × 480 VGA grey-scale capability and a speaker. Input is multiple, and includes a keyboard, a glide-point tracking device, a microphone, and a single shot video camera. Software includes voice recognition software, and Windows 95 as an operating system. Compared to Figure 7.1, every device has indeed shrunk to the size of the head of a pin, while sacrificing none of the capability of the system in Figure 7.1, and gaining several new capabilities besides.

The capabilities of this system are to make augmented memory possible. In augmented human–computer interactions, unlike the alternative virtual environments, users retain their normal suite of perceptual multisensory (and therefore navigational and mobility) inputs, but the computer's I/O functions enter into the multisensory realm. The user interface for the University of Oregon system can take touch, voice, video image, or keyboard input, and gives sound, image or text output. The suitability of these senses for GIS outside of the visual context has been discussed by Shepherd (1994). Augmented reality, in which the user can be partially immersed in a virtual reality, but only to supplement the sensory input, is the basis for the scenario at the start of this chapter. While virtual display of 'total immersion' virtual environments is an active field of study (the so-called virtual reality: see Batty et al, this volume), so far scientific applications have been limited to flight simulation and training, and psychological experiments in human spatial cognition and navigation (Loomis et al 1993). Cartwright (1997) has noted that 'knowledge-building and decision-making could be made in an artificial world completely devoid of nature's checks and balances'. Augmented memory and augmented reality offer much more as far as the geocomputational future is concerned.

Hardware development at the research frontier is difficult to review, and the essential path of scientific peer review, careful testing and replication will be necessary before many of the new hardware systems will become generally available. Nevertheless, technology has the apparent capability of moving at the speed of light, and the following section will review part of the suite of hardware components which, when assembled together, could make the opening scenario a reality. In following the traditional approach to computer hardware, I will divide the systems into input and output devices. As will become clear, such a division is as fuzzy as that between hyperinteractivity and mobility. In any case, they both almost fit on the head of a pin.

Shepherd (1994) has suggested that GIS users 'would benefit in many ways from being able to exchange information with the computer using multisensory GIS'. In his paper, he divided the input and output capabilities of GIS by their suitability for use with the human senses (Table 7.1). As input, most favoured were sight and hearing, though sounds, gesture, touch, speech and force were also feasible. A review of input to wearable computers has also been made by Stamer (1995), used

Table 7.1 *Shepherd's potential uses of human senses in GIS. Source: Shepherd (1994)*

	Human input to GIS	Human output from GIS
Sight	○○○○○	○○
Speech	○○	○○○
Touch	○○○	○○○
Force	○○○	○○○
Gesture	○	○○
Hearing	○○○○○	N/A
Smell	○	N/A
Taste	○	N/A
Sound	○○	N/A

Note: Numbers of circles indicate the strength of the role

to inform the building of an off-the-shelf wearable system ('Lizzie') by the Wearable Computing Group at the Massachusetts Institute of Technology (MIT; Figure 7.2). The principal division of the hardware review was by the form of human input, and this division is also followed here (see Table 7.1).

7.3   Ubiquitous Computing Input Devices

7.3.1   Speech

Speech input is predominantly via specialised tunable, noise-cancelling microphones. These are worn close to the mouth on a headset. Speech recognition software capable of vocabularies of thousands of words is now readily available, with neural-net-based training capabilities that learn (sometimes assisted by typing) the speech patterns and accent of the user. The IBM Voicetype software on my home computer cost less than $100. This means, of course, that unless the software can develop separate profiles for different people, then a trained voice recognition system is as personal as, well, clothing! Stamer points out that voice systems are not suitable for privacy or for stream-input, since they require constant editing of the flow. Tape storage of voice input for voice recognition purposes would be possible, but would not be truly interactive.

7.3.2   Handwriting

Handwriting recognition followed the development of the Personal Digital Assistant (PDA) industry, led by Apple's Newton PDA. The user writes on the screen of the system, or on a separate pad, and the machine attempts to use image processing methods and look-up techniques to match pen-strokes to characters. While successfully implemented by the Post Office to read zip (post) codes, street addresses and so forth at very high speeds, most other systems are slow (30 words per minute maximum: Stamer 1995), and the need to write on a surface is an impediment to mobility. The screens are also small, and the need to carry a pen for the interaction also detracts from full mobility.

7.3.3 Gesture and Body Motion

Although not considered in detail by Stamer, the flexibility of the wired glove as an input device is remarkable. Several different designs and technologies have emerged, pioneered in 1989 by Mattel's inexpensive Powerglove for the Nintendo video game. Glove systems can be quite sophisticated. While the Powerglove senses motion by signals to two microphones which must be placed around it, gloves with multiple sensors and miniature accelerometers can follow complex hand movements and gestures. In virtual reality immersive environments, this has allowed hand regulation motion control and even 'picking' of objects, including tactile feedback to allow grasping and touch (Kessler et al 1995). Within the last two years, primarily in the movie and animation entertainment industries, there has been developed a suite of full body motion tracking technologies. Most require motion to be constrained mechanically, or use laser reflectors on the body in a space containing sensors. In some cases, large numbers of signals are possible. One set of technologies track markers or reflectors on the human, either on all limbs or at many places on the face for expression. Alternatively, this can be done mechanically, by a device called a 'Waldo'. These devices are very cumbersome, yet allow extremely fine motor control input.

7.3.4 Keyboards

Keyboards are highly flexible input devices. Skilled courtroom stenographers can achieve 300 words per minute (Stamer 1995). QWERTY keyboards, actually originally designed to slow down typing for mechanical typewriters, can achieve 240 words per minute. The major problem for mobile and wearable computing is size. There is a practical limit to how small a keyboard can be and still remain useful. Some wearable computing applications have used shortened keyboards strapped to the forearm, so that typing is single handed. Another promising alternative is the Twiddler (Figure 7.3). This hand-held keyboard uses multiple shift keys and sequences to include a full character set, and is very flexible in use. It includes a small tracking device to move a regular mouse and has been adopted by many of the universities conducting research on wearable computing. Drivers for several operating systems are now available, including Linux and Windows. Several alternatives are shown at *http://lcs.www.media.mit.edu/projects/wearables/keyboards.html*.

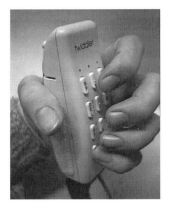

Figure 7.3   *The Twiddler keyboard. This is a full character set keyboard useable in one hand for mobile and wearable systems. Courtesy of Handykey Corporation (http://www.handykey.com), 141 Mt Sinai Avenue, Mt Sinai, NY 11766. Used with permission*

Figure 7.4   *Immersion Corporation's FEELit mouse, with force tactile feedback. Immersion Corporation, 2158 Paragon Drive, San Jose, CA 95131. Web Site: http://www.immerse.com. Used with permission*

7.3.5   Tactile

One of the most common input mechanisms that can provide tactile feedback is the joystick. Moving a joystick can encounter a resistance force that can be integrated with shaking to simulate sensations. 'Stick-shaker' tactile response is very common in the flight simulation industry, and many derivative products are common in the video game industry. An interesting development in tactile control is the FEELit Mouse (Figure 7.4). This ordinary mouse uses force feedback to the hand to make familiar parts of the GUI, such as scroll bars and window edges 'feel' as though they have different textures, cross-sections and heights. Adding full tactile feedback to a glove would be a little more difficult, but could be done in ways other than actual contact with the object involved.

7.3.6   Vision Tracking

Vision tracking uses many technologies to measure the point on an image when the user's eyes are focused. A first technology is to measure the location of a light beam emitted from a head-mounted device as it hits the screen. A second is to record the minor differences in pressure on the surface of the skin as the eyeballs move. A third is to place a camera facing the users' eyes, and to compute the vision focus by image processing and triangulation. Another method, common in flight simulators and virtual displays, is to sense head motion using accelerometers mounted in a helmet. For interactive use, 'clicking' on a location is achieved by blinking the eyes or by dwelling on one spot with one's glaze. Costs of these technologies have dropped significantly in recent years. Many, but not all, wearable systems have incorporated one or another of these devices, and use it for most input to the system.

In addition, most wearable systems include either a single frame or video camera. This allows images to be acquired, stored and transmitted. Work at the University of Toronto has allowed the linking of such a single-frame camera to the World Wide Web, allowing anyone on the Web to 'see through' the user's eyes.

7.3.7   Direct Neural

Direct neural input systems literally read one's thoughts. They take advantage of the various ways in which human movements and reactions have detectable components elsewhere within the metabolism. The term bionics is often used to cover eye movement and the array of alternatives. These include monitoring emotion (blood pressure and heart rate, dilation of the pupil, etc.), monitoring facial expressions and lip reading, and probing the brain. The latter is possible in several ways, including the use of the electroencephalogram (EEG), which are measures of surface electrical signals on the scalp reflecting brain signals and measuring functional brain wave topography. In each of these cases, the input mechanism is far from direct. A significant amount of training is required for the user to achieve control over the pointer. Vidal (1997) has argued that these technologies are still to be developed to the point at which they are useable. Stamer (1995) has suggested that the accomplishment of direct thought input is largely impossible without direct invasion of the brain by probes or electrodes.

7.3.8   The Global Positioning System

The US Global Positioning System, since being declared operational in 1995, has provided the opportunity to add direct input of positional coordinates to GIS and other applications. The system consists of 21 satellites and three active spares in six orbital planes with 55 degree inclinations and 12 hour period orbits at 20 200 km altitude. A flurry of recent technical advances has led to several key improvements in the positional accuracy of GPS, to the extent that positional accuracy now approaches or exceeds the positional accuracy of even large-scale maps. The main GPS signals are in the microwave, an L1 carrier at 1575.42 MHz and an L2 carrier at 1227.6 MHz, which carry the Coarse Acquisition (C/A) code and the encrypted P-code, with accompanying ephemeris data. These signals presently have several imposed restrictions that determine accuracy, including an anti-spoofing measure and random-number-based signal degeneration through Selective Availability. Nevertheless, the use of differential processing from one fixed and one mobile unit,

Figure 7.5 *GPS field applications. Left: the GARMIN digital GPS mapping system, which maps the GPS input onto a digital highway map anywhere in the US. Right: the Laser Technology Inc. Geolaser integrated differential GPS and laser range finder. Sources: http://www.garmin.com and http://www.lasertech.com/laserproducts/geolaser.html. Used with permission*

or use of corrections broadcast in any of several ways, can improve accuracy from the standard C/A accuracy of 100 m horizontal and 156 m vertical to carrier phase tracking with accuracies in the centimetre range. This has opened up many applications in vehicle guidance, personal navigation, fleet tracking and surveying. Real-time differential measurement from large numbers of satellites is now common, and is even possible in highly mobile systems. Some recent technical applications have included embedding GPS into a national digital map display, into laser ranging and bearing devices, and into cameras that record GPS data directly onto exposed film (Figure 7.5).

There are many drawbacks to GPS, even in standard navigation systems. Signal accuracy diminishes because of variations in the positional elements of the satellite constellation, because of screening by buildings and vegetation, or because of atmospheric effects. The system will also not work indoors, or in areas where parts of the horizon are screened out, such as canyons, streets and valleys. Nevertheless, the system allows repeated position tracking (less that a second per fix) in a large variety of locations, and is a critical field data collection element for ubiquitous computing systems. While slow to be integrated into wearable computers, GPS has significant potential in this respect. In addition, GPS is an essential and critical link to geographical space, and the only way to relate mobile computing units together on the face of the Earth.

7.4 Ubiquitous Computing Output Devices

Output capabilities of wearable systems almost all include a display of some kind. Sound is also used, with an earphone speaker often integrated into a headset or head-mounted display (HMD). Sound output can be used quite effectively for many purposes, but suffers from a lack of personal privacy and also interrupts the user's attention. Tactile output is hard to separate from input, since the human–machine interaction is two-way. No channels for the more exotic means of output exist (smell, for example), although some forms of computing, such as Braille output and images of hand gestures for sign language are used in special-purpose applications. Output, then, largely means the visual display. Getting the computer screen into the human vision field has been an experimental venture. Some recent breakthroughs, however, have made great leaps forward.

Many of the first generation of wearable computing systems took advantage of liquid crystal displays (LCDs), but were capable of only gray-scale display at low resolution. Colour and higher resolution came about using light emitting diodes (LEDs), which can place a small, but reasonably high-resolution screen directly into the human vision field. The wearable computing systems shown in Figure 7.2 use a visual display that places an LED into the vision field. This system is remarkably effective, simulating the visibility of a 20 inch (50 cm) monitor at a usual visual distance of around 24 inches (60 cm). Reflection Technology's P5 display is commonly used (see: *http://www.reflection.com/p5disp.html*). This is a monochrome display with crisp red and black images at resolutions up to 864 × 256 using an LED and a mechanical scanner. Pixels have a 1:2 aspect ratio, refresh at 60 Hz like a television set, and the display weighs only 32 g and consumes only 191 mW of power.

Whether one or both eyes is used is important, because single-eye views preclude the use of simulated stereo and limit visibility to one dimension, the flat surface of the display. Some early augmented memory applications have included face recognition. The forward-looking video camera scans a frame including a human face, and software then uses image processing to match key structural features, such as the tip of the nose, and the edges of the eyes, and a database of faces is queried to retrieve the match and display the name. Given the difficulty of 'putting a name to the face', this is in itself a powerful application. Full use of the vision field allows a 3-D database to be overlain and annotated in visual perspective. City streets, for example, could be annotated with their names and the direction of traffic flow.

A significant recent development has been miniaturised LEDs small enough for inclusion in or onto the glass of ordinary glasses. One person in the left image of Figure 7.2 is wearing such lenses. These glasses are manufactured by the Micro-Optical Corporation of Boston, Massachusetts (Spitzer et al 1997; Grossman 1998). While still experimental, this technology shows great promise for bright colour images at high spectral and spatial resolution. The company MicroDisplay in Cambridge, Massachusetts, has introduced a prototype display device produced at MIT's Artificial Intelligence Laboratory with DARPA funding. The new technology combined liquid crystal displays with silicon fabrication technology to generate extremely high-resolution small displays. For example, Figure 7.6 shows a display device that is one square centimetre in size, yet has a 2000 x 2000 display. Colours are created by passing white light through a triad of microgratings to generate diffraction, with no light loss and extremely low power consumption.

Display hardware based on this technology is easily capable of direct integration into normal glasses and contact lenses, and so offers perhaps the best future for wearable ubiquitous computing systems. A survey of available technologies for wearable computer displays is located at *http://lcs.www.media.mit.edu/projects/wearables/display.html*. Also of interest is the wearable computing timeline at *http://lcs.www.media.mit.edu/projects/wearables/timeline.html*. Finally, to conclude the discussion of displays, one company (Microvision of Seattle, Washington: see *http://*

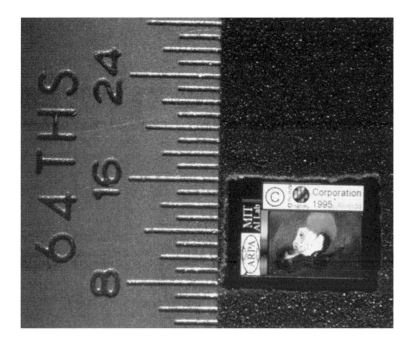

Figure 7.6 *A less than 5 mm diagonal colour display with 5 micron pixels based on LCD and silicon wafer technology. Image courtesy of MicroDisplay Inc.* (http://www.microdisplay.com/index.html). *Used with permission*

www.mvis.com) is developing a laser technology to draw computer images directly onto the human retina, using what the company calls a virtual retinal display. The company anticipates applications in defence, healthcare, industry, wireless communications, and consumer electronics. It seems likely that at least one or other of these two technologies will become commonplace for use in highly mobile computing.

7.5   Hardware Drives Software Drives Research

This chapter began with the view that, in the early phase of GIS, the peripheral hardware of graphics devices dominated the science around GIS. This was followed by the era of developing software interoperability, in which functionality became independent of device, operating system, and platform. This brief survey has shown that ubiquitous computing, high-mobility systems and the demands of hyper-interactive user interfaces will once again bring the computational hardware back to the forefront in GIS and cartographic visualisation research. The same fundamental computational architecture remains, but miniaturisation is now making ubiquitous computing possible.

It is worth a brief diversion to discuss what areas of computing can now be delivered to the eyeball, ear, and twiddler! First and foremost, the stand-alone power of the microprocessor by itself, when coupled with memory, is significant. Many applications involving augmented memory and augmented reality can use pre-stored information. Image processing, standard GIS functionality, stored GIS and modelling results, and databases could all have significant applications if available for field computing. If a network link is added, many more applications become possible, including those in the scenario at the beginning of the chapter. A network connection to a ubiquitous or wearable system at the least requires cellular or other communication connection (itself a highly mobile technology), the ability to download images and data, to browse functionality, and two-way video and stationary image communications. Most of these are existing network functions on the Internet and the World Wide Web. There is little difference between a Web-cam mounted on a static site or one moving through space (Cartwright 1997). With the whole Internet available, the body of material deliverable to the field user (data and information) becomes significant.

Downloading digital map data on demand, searching out individuals to contact by email, voice-mail or through two-way video links, even retrieving real-time views from static and overhead (e.g. satellite) cameras becomes possible. Wearable computer users could be walking down the street and simultaneously watch themselves walking both from the other end of the street and from above. One could almost literally have eyes in the back of one's head, and in many other places besides! Imagine introducing students to the concept of the Universal Day and time zones by looking in on a Web-cam focused on a clock in every time zone around the world. Such Net-based experience makes possible new 'exploration' activities. Virtual travel and remote experience, such as the recent Mars lander, are augmented geography. As with the child staring at the pages of an atlas, they enhance the geographical imagination. While the entertainment and travel industries will (and have already) been early exploiters of the technology, the possibilities for direct navigation, even beyond the in-vehicle system, abound. Furthermore, many new capabilities can be provided for those unable to travel, such as the vision and mobility impaired.

Computers have had far wider scientific impacts than computation and database distribution alone. Among the most significant have been simulation and visualisation. To allow the creation of a data world, that can be explored by visual browsing or by experience (as in simulation), opens up manifold new applications and capabilities. Astronauts and pilots are often heard to say that their potentially dangerous task was 'just like the simulations'. There can be little doubt that mechanical and now fully digital simulation has led to whole new industries in systems design and testing. Adding simulation to human experience has many possibilities. A virtual traveller could try several paths to a destination before setting out. A cartographer could view hundreds of permutations of a map in context before choosing the best design for a specific task. A student anxious about crossing the campus at night could practice the task many times in varying lighting conditions to choose the safest path. One common element in these simulations is that not one, but many visualisations could hold the geographical answer to a problem. Multiple realisations have been used to show the degree of uncertainty present in data using stochastic imaging (Journel 1996). Many applications can now generate not a single model or GIS output, but the full set of possibilities within a set of ranges or tolerances. Having not only the map but the uncertainty map, and being able to use it in real time for specific decisions, may be the most valuable contribution of ubiquitous computing.

Such an approach might be termed 'intelligent choice'. This decision-making system goes far beyond the notions of educated guess or even informed choice. Navigators using augmented reality could not only know with certainty that a particular walking route will reach their destination, and in what length of time, but by browsing Web or overhead cameras for the whole route, any navigational uncertainty can be removed and the choice updated in real time. Where to cross the road with minimum risk or to avoid a puddle, whether or not the mugger (visible and identified) two blocks away could reach them by any possible route before they get home, identifying where the nearest store is located on their route that has fresh flowers for a birthday surprise (advertised on the Web), perhaps resequencing or rearranging the order of several errands, all are examples of intelligent rather than informed decision-making. For geography, the very discipline centre can return to the field itself. Just as the recent international mountain and biking exploits can be followed by schoolchildren on the Internet anywhere, so any geographer will be able to share the field exploration and discovery of any other.

7.6 Research Challenges for the Geocomputational Future

As with all science, glaring gaps in the conceptual knowledge necessary to support the vision of ubiquitous computing abound. First and foremost, with an entirely different technology, just what should be displayed? GIS has allowed a layered concept for map information, and the idea of switching information layers on and off, and using layers in combination with analysis to derive outcomes, is commonplace. Now the data, and the outcomes, even the analysis, can be delivered into the very environment in which the maps are used. The wearable system user can be the data gatherer, analyst and user all at the same time. The potential for real-time data update is immense. For example, the advent of cellular communications and car phones made it possible for an individual experiencing a traffic hold-up or an accident to report the incident for broadcast information sharing. Obviously this will happen with ubiquitous computing. Reporting incidents, map errors, prediction and simulation errors or successes, and many other activities becomes possible. The reporting can be global, instantaneous, and fully distributed.

The next issue facing research and related to display is what can be done about the problem of information overload. A driver staring at his virtual street map may miss a turn, or worse. A student viewing 12 software windows each showing a prominent university geography professor debating a concept might get nothing out of the experience. In short, what are the limits of perceptual input to the human reasoning system? How much memory do people want to augment their real memory with, and when does augmented reality interfere with reality and normal human functions? What is information overload? Can it be reduced, eliminated, or minimised? Is overload temporal, constant, transient? Does it vary by individual, by intellect, by gender? Does information overload lead to partial system use or to 'tuning out'? Can we turn on the manual override and still retain some of the same benefits? And finally, if mobile computing systems provide rich geographical and spatial information to novice users, how can such users be made sufficiently spatially literate to make effective use of the innovations?

Another role is largely societal. The information age has turned many traditional concepts of economics on their heads. One can become a market leader by giving away a product, or by staying small and divided rather than by growing. Given that many controls on information are market driven, and that a free data framework appears necessary at public expense, where and how should information controls and filters be applied? Obviously, wearable computing reeks of spies, personal surveillance and big brother watching us. If local police departments can hook up and monitor cameras at street intersections to watch traffic, or a security service wire a building, why shouldn't any ubiquitous geocomputing user have access to the signals? If television networks and courts can get access to this data, why should not the public? Who and what mechanism will prevent abuse of the systems and the knowledge they can create? The same camera that provides my security while I walk down the street at night also violates my privacy if it reveals my identity and position. Geographers probably cannot answer these questions, and arguably are not even those best equipped to try. Nevertheless, parts of the discipline have shown great interest in these issues, and their study.

At the practical level, many of the future research issues for ubiquitous computing are well represented in the current research agenda, with one exception. As far as the problems of interoperability, standards, Web-based mapping, access, bandwidth and system efficiency are concerned, progress in research is already under way. The exception is that of the user interface. Current research is oriented toward either extension of the existing GUI models (Egenhofer and Richards 1993) or towards an entirely different model of interaction, the wall-board of the smart room approach. On the other hand, an encouraging line of research is working towards the software realm from the context of studying the means by which humans interact with their perceived or immediate behavioural spatial environment (Egenhofer and Golledge 1998). Careful merging of this research realm with software design could lead to some effective user interfaces for ubiquitous systems. To date, the prototype wearable systems have used Windows, Linux or even DOS as their operating systems. Obviously the smaller operating system kernels, and those that emphasise port connectivity, network connections and embeddable functionality, such as UNIX, will prosper in the wearable environment. At the moment, research has barely begun to examine the interoperable miniaturised operating system problems.

Finally, taken to extremes, the future of geocomputation offers some both exciting and frightening prospects. Vidal (1997) has suggested that human emotional inputs are also useable in wearable and ubiquitous computing systems. While it would be of great benefit for a doctor to be able to remotely read a patient's essential signs such as pulse and heart rate, the potential for abuse with sentient input (and output?) is immense. On the other hand, for astronauts and pilots, it is an essential ingredient. With field computing, could overexertion, altitude sickness, exhaustion, heatstroke, hypothermia and the other traditional enemies of fieldwork in extreme environments be controllable or managed? And what will happen when to conduct such monitoring, the hardware of wearable computing becomes invasive? Can computers become surgical implants, like the pacemaker? Can the retina really be used directly for input and output? Can the various sources of measurable human brain waves really be used as inputs for wearable systems? With cyberspace bionics moving from science fiction to practicality, led by ubiquitous computing, a strange and powerful technology, industry, and future offers itself. Personally, remembering that the first volunteer mechanical heart patients broke new ground by simply surviving from day to day, I will wait until the bugs are all out before having ROM and LED displays surgically implanted into my head. The pioneers of cyberspace bionics beyond the wearable will be pioneers indeed!

A final aspect of the fully distributed ubiquitous geocomputational future is the range of modes in which such systems could operate. At one extreme, a geo-computational cybersociety could consist of watchers. All could be voyeurs from any location, and all might have the same data access. At the other extreme, individuals could have superior powers to be actors rather than watchers. As the action capability increased, the user could first read single input source signs (video, sound, heart rate, breathing rate, body temperature) and monitor them. As the level increased, the user could activate sensors (turn on an overhead camera, begin identifying people in the visual field) of another source. Finally, full action would actually manipulate either robot or human actions at the source (drive a car on Mars or even on Earth, tell a child to come home from the park, turn on a measurement instrument). Our existing primitive networks with only limited active/passive control have already brought with them their own differential ownership, geographical divisions, protocol and etiquette. The World Wide Web is largely a passive, watcher system, like television. A ubiquitous geocomputational future of field computing begs action and movement, and forces geographical theory into daily practice. New challenges and potentials are as ubiquitous as the computational environment that is now evolving so rapidly.

7.7 Conclusion

Ubiquitous geocomputation is a new suite of technologies with immense promise for geography. The technologies offer to advance the promise of GIS by eliminating the spatial separation of the peripheral devices for computing by miniaturisation of input and output devices. So radical is this divergence of path from that of traditional geocomputation that a new hardware era will be the initial scientific consequence. Once more, technology will drive science, and it will be years before the geographical consequences of wearable and ubiquitous computing become evident, and the critical geographic information science questions emerge. In the best of all possible worlds, the applications possible and new science questions solvable promise a new age of discovery for geography. In the worst of all possible worlds, the potential for abuse, for loss of privacy, and for alteration and manipulation of the human consciousness also threaten. I dread the day when I am woken from a sound sleep by a noisy, flashing advertisement projected onto my retina urging me to download a new free Web-browser, one that I cannot turn off without mentally focusing on a dark grey 'Decline' button hovering at the far range of my peripheral vision. Nevertheless, as a professor, I look forward to teaching students directly while walking in the mountains or on the beach, and to sitting again at a desk devoid of a massive, noisily fan-cooled, cathode ray tube. And were I the wounded passenger in the plane lost in the mountains, I might even tolerate the pain of surgical implants, perhaps even the noisy advertisements, to have my rescuers accomplish an apparent miracle.

References

Auerbach 1972 *Auerbach on digital plotters and image digitizers.* Auerbach: Princeton, Auerbach Publishers

Cartwright W 1997 New media and their application to the production of map products. *Computers and Geosciences* 23: 447–56

Clarke K 1995 *Analytical and computer cartography.* Englewood Cliffs, Prentice-Hall: Chapter 2

Clarke K 1997 *Getting started with geographic information systems.* Upper Saddle River, Prentice-Hall: Chapter 10

Coppock J T, Rhind D W 1991 The history of GIS. In Maguire D J, Goodchild M F, Rhind D W (eds) *Geographical information systems: principles and applications.* Harlow, Longman/New York, John Wiley, 1: 21–43

Egenhofer M J, Golledge R G 1998 *Spatial and temporal reasoning in geographic information systems.* Oxford, Oxford University Press

Egenhofer M J, Richards J R 1993 Exploratory access to geographic data based on the map overlay metaphor. *Journal of Visual Languages and Computing* 4: 105–25

Foresman T M (ed.) 1998 *The history of geographic information systems: perspectives from the pioneers.* Upper Saddle River, Prentice-Hall

Forman P 1987 Behind quantum electronics: national security as a basis for physical research in the United States 1940–1960. *Historical Studies in the Physical and Biological Sciences* 18: 199–229

Grossman W M 1998 Wearing your computer. *Scientific American* 278: 1

Journel A G 1996 Modelling uncertainty and spatial dependence: stochastic imaging. *International Journal of Geographical Information Systems* 10: 517–22

Kessler G D, Hodges L F, Walker N 1995 Evaluation of the CyberGlove™ as a whole hand input device. *ACM Transactions on Computer–Human Interaction* 2: 263–83

Krygiel A 1997 *The US defense vision and its implications for GIS technology.* Paper presented at InterOp97 International Conference and Workshop on Interoperating Geographic Information Systems December 3–4 1997, Santa Barbara, *http://www.ncgia.ucsb.edu/conf/ interop97/*

Loomis J M, Klatzky R L, Golledge R G, Cicinelli J G, Pellegrino J W, Fry P A 1993 Nonvisual navigation by blind and sighted: assessment of path integration ability. *Journal of Experimental Psychology: General* 122: 73–92

Negroponte N 1995 *Being digital* New York, Knopf

Shepherd I 1994 Multi-sensory GIS: mapping out the research frontier. In Waugh T C, Healey R C (eds) *Advances in GIS research Volume 1.* London, Taylor & Francis: 356–90

Spitzer M, Rensing N, McClelland R, Aquilino P 1997 Eyeglass-based systems for wearable computing. *Proceedings International Conference on Wearable Computing.* Cambridge

(MA) October 13–14th 1997: *http://mime1marcgatechedu/wearcon/graphical_wearcon/wearcong.html*

Stamer T 1995 *Input devices: there is no Holy Grail. http://lcs.www.media.mit.edu/projects/wearable/input-guidelines.html*

Tobler W R 1959 Automation and cartography. *The Geographical Review* 49: 526–34

Vidal J J 1997 Cyberspace bionics. Sequel to 'Toward direct brain–computer communication'. In Mullins L J (ed.) *Annual review of biophysics and bioengineering.* Palo Alto, Annual Reviews Inc Palo Alto: 157–180 (draft available at: *http://wwwsv1u-aizuacjp/~vidal/cog.html*)

Weiser M 1991 The computer for the twenty-first century. *Scientific American* 265: 94–102

8

Modelling Virtual Environments

Michael Batty, Martin Dodge, Simon Doyle and Andy Smith

Summary

In this chapter, we explore the way in which virtual reality (VR) systems are being broadened to encompass a wide array of virtual worlds, many of which have immediate applicability to understanding urban issues through geocomputation. We sketch distinctions between immersive, semi-immersive and remote environments in which single and multiple users interact in a variety of ways. We show how such environments might be modelled in terms of ways of navigating, processes of decision-making which link users to one another, analytic functions that users have to make sense of the environment, and functions through which users can manipulate, change, or design their world. We illustrate these ideas using four exemplars that we have under construction: a multi-user Internet GIS for London with extensive links to 3-D, video, text and related media, an exploration of optimal retail location using a semi-immersive visualisation in which experts can explore such problems, a virtual urban world in which remote users as avatars can manipulate urban designs, and an approach to simulating such virtual worlds through morphological modelling based on the digital record of the entire decision-making process through which such worlds are built.

8.1 Definitions

Virtual environments are digital simulations of situations, real or fictional, in which users are able to participate. Participation and the way it is achieved are the key components which make the environment virtual, and it is generally agreed that such participation must be engendered so that users are able to feel that they are 'present' within the environment and are able to interact with the simulation if only by navigating and moving within the scene. These kinds of environment might at first sight appear very specialised although, at one level, the use of any computer

---

Geocomputation: A Primer. Edited by Paul A Longley, Sue M Brooks, Rachael McDonnell and Bill Macmillan.
© 1998 John Wiley & Sons Ltd.

software embodies elements of virtuality. As software has become ever more graphic and as users have begun to interact with software using the icons of point and click, there is a sense in which all computation is taking on elements of virtual reality (VR). Predictions for the way we are likely to interact with machines in the near future suggest that VR will become the dominant form of such interaction (Negroponte 1995).

The way VR has developed is instructive for understanding what is now possible in constructing virtual environments which embody geocomputation. The first steps towards VR involved building digital simulations as close as possible to single users, 'immersing' them within the environment in such a way that they were connected directly, through peripheral devices such as headsets, data-gloves and the like (Clarke, this volume; Heim 1997). These developed in parallel with aircraft simulators where the emphasis was on creating realistic graphic environments in which users could execute tasks as close as possible to those conducted in a real situation. Until quite recently, the emphasis in VR has been on three related issues: representing environments using 2-D and 3-D computer graphics in such a way that users can navigate within them, introducing tactile and olfactory senses into such environments as well as achieving graphic realism, and concentrating on the way such environments are used by single individuals (Rheingold 1991). As recently as five years ago, Kalawsky (1993) noted that traditional VR was divided into the kind of immersive activity that developed to embrace various human senses, and desktop VR based on animated computer-aided design (CAD). Both activities were dominated by the direct viewing of virtual environments rather than by processes for undertaking non-visual activities within visual environments.

However, there have been very rapid changes in the concept of virtual reality since Kalawsky (1993) charted the field. Other developments in computing, specifically scientific visualisation, have embraced VR, while the movement of computation to the Net is changing the idea of participation and presence within virtual environments. To make sense of all this, it is worth developing a fairly formal approach which identifies the key components of VR so that changes due to new technologies can be easily understood. Virtual environments bring two domains together: the *digital environment* and *users* who participate in some way within such environments. Both the environment and users must be represented and of course connected together. Environment and users interact and, within such interaction, various behaviours can be modelled. Moreover, the environment itself might be modelled independently of users and vice versa, although to effect virtuality these processes must come together.

A threefold paradigm of *representation, modelling* and *connection* is central to the construction of virtual environments. So far, most images of VR emphasise the environment, usually represented in 3-D form using state-of-the-art graphics and CAD, with users being connected to such environments through protocols – models – which enable them to sense and perhaps manipulate the environment through navigation and other forms of body movement. However, using this threefold paradigm in the context of dramatically changing technologies opens up many new possibilities for VR. For example, representation of the environment need not be restricted to 3-D; it might be any-D, from the 2-D map paradigm to its generalisation in mathematical space. Modelling is then not merely navigation or body movement but might involve any behaviour pattern one might care to think about, from the activity of 'doing science', to 'manipulating financial transactions', through to 'taking the dog for a walk' in a virtual world. Connection, however, is the most rapidly changing aspect of VR at present. Fully immersive VR, from whence the activity emerged, is now complemented by various kinds of semi-immersion in which users are able to combine the normal material world of debate and discussion with interaction in a virtual world. VR theatres and CAVES are the most obvious portrayals. ('CAVE' is an acronym for 'CAVE automatic virtual environment', a recursive definition that evokes thoughts of 'Plato's Cave': Cruz-Neira et al 1993).

However, the semi-immersive nature of these environments, where interaction with the virtual world is embedded with interactions in the material world, is not the main focus *per se*. Once fully immersive VR is opened to the material world, not *single* but *multiple* users can participate. In a sense, interacting alone in any environment will always be a special case. Facilitating more than one person to interact enables a full range of social interaction to be represented and modelled. In developments of VR technology into theatres and CAVES, interaction is still direct, but what is now happening as computation drifts out into the ether is that presence and connectivity are becoming *remote*. Users can access virtual environments across networks, can be represented graphically as 'avatars', a term used to define a virtual representation of an animate object (from the Hindu definition of the 'descent to Earth of a deity in human, animal or superhuman form' – OED 1991). They can then engage with others who are connected at different remote locations with respect to some task. These kinds of networked virtual worlds are growing very rapidly at present, providing environments in which many new kinds of representation and modelling are able to take place. The next step, already happening, is linking such remote interaction with semi-immersive and immersive VR, thus generating a veritable cornucopia of virtuality.

We will first begin to classify virtual environments with a specific focus on geocomputation. This will frame the scope of the examples that we will illustrate in this chapter, but before we select these, we will describe a variety of different representational and modelling techniques that have and can be used in constructing virtual worlds. We will then focus on four examples from the wide array of possibilities that we chart, concentrating on environments in which the three key activities of information, analysis and design are combined into four types of geocomputation. We will begin with information-based analysis based on Internet GIS with various multimedia, which opens up the world of GIS to a much wider set of participants. We will then illustrate how conventional geographical (spatial) interaction models can be explored virtually, indicating how optimal locations can be investigated visually by many experts. This is followed by an illustration of how group problem-solving and the design of real urban environments can be achieved remotely through networked virtual worlds which mirror the real world in question. And finally we will show how fictional virtual worlds, designed with no specific purpose in mind other than social interaction, might form rich datasets for studying how real urban environments might be explained and simulated.

8.2 Classifying Virtual Environments

Providing maps for any rapidly changing emergent field is essential but impossible. Virtual reality has no formal definition, as we have already implied. In one sense it might be seen as the emerging interface to digital computation *per se* but our distinction between environments and users and the threefold paradigm of representation, modelling and connection does provide a convenient starting point. It is immediately obvious that all non-digital environments have some counterpart in the digital world, no matter how realistic, and although most of these have not been developed, nor may they ever be, then this still points to a vast array of virtual environments that might someday be realised. In terms of the environments to be constructed, it is worth making a distinction between *real* and *fictional* worlds. The line between such worlds is inevitably blurred; it is hard to define any world which is simply one or the other and this merely recognises the old adage that 'truth is often stranger than fiction'. But this is important because much of the technology and progress in VR comes from fictional applications in games or in simulated situations which have no link whatsoever to any material reality.

There is another important distinction between *spatial* and *non-spatial* environments. There is a further distinction between *spatial* and *aspatial* (after Webber's (1964) place and non-place urban realms) but here all we need be concerned with is environments which can be represented in 2-D or 3-D and those that do not bear any resemblance to such Euclidean space such as those in mathematics, those based on writing and verbal interaction, those based on sound, etc. Most virtual environments to date, particularly those based on full immersion, are spatial, usually 3-D, and base representation on the material world of buildings, landscapes and interiors. Part of our quest here is to show that these kinds of world are likely to be the exception rather than the rule as VR continues to develop (Kitchen 1998). As with reality and fiction, spatial and non-spatial blur into one another and many applications are likely to contain elements of each.

First applications in VR consisted of *single users* being immersed in digital environments, and this has been reinforced time and again in the last 20 years as the desktop paradigm has come to dominate computation in general. In fact, immersive VR has diffused quite rapidly, particularly in the games industry where the personalised headset has become *de rigueur*. However, with the growth of semi-immersive and remote VR, the idea of *more than one user* being part of any virtual environment has developed very fast (Batty 1997a). Indeed, it is likely that virtual environments will eventually be defined as environments which enable more than one user to participate. Of course, the social structures that these users adopt will depend upon the reasons for interacting through such environments. Any social group in the material world might interact in the virtual but here we will define four distinct types: those using VR for science, those using VR for some professional task such as design, or problem-solving, those using VR for political purposes, and those using it for leisure pursuits.

It is difficult to develop this classification of users further without referring to the purposes which VR is likely to serve. We will make the common distinction between using VR for *information* and advice, for *science* (analysis), and for *design*.

All these tasks characterise some form of serious and considered directed inquiry although we must also identify leisure pursuits as having an important influence on the development of VR. Relevant to use is the nature of that use, the common distinction being between (often frequent) routine versus strategic (one-off) usage. Most applications to date are one-off, although once VR really begins to disseminate through diverse applications this will change. VR is more likely to become routine during the next decade, just as GIS and CAD are becoming routine, due largely to the fact that the elements of representing and modelling virtual environments are quite basic functions in themselves, do not involve esoteric science, and are fundamental to interacting with any kind of computation.

As we have already stated, the key activities in VR are *representation, modelling* and *connection*. *Representation* of environments and users can be *visual* or *non-visual* or some combination thereof. Visual representations can be of real-material systems – buildings, landscapes, interiors – or of abstract forms such as those associated with the analysis of problems in mathematical or statistical space. 2-D map-based representations lie between the real and the abstract but, as we implied earlier, the lines between these categories are always blurred. Non-visual representations involve text, sound and numbers whose form can be visual but whose intrinsic representation is not. Mixtures of both visual and non-visual characterise all environments and users. The most visual environment–user system involves users appearing as avatars in some 3-D space within which they can see themselves and others navigate. But systems based on pure text or even sound, such as Internet relay chat (IRC), comprise VR, for users are often imbued with a strong sense of presence, such environments meeting all the requirements for VR posed above.

Modelling virtual environments include several processes which simulate the way users and environments respond to each other; we define modelling as the shorthand for ways of interacting with the environment according to standard behavioural rules applicable to particular users. *Exploration* of the environment is important and this involves some mode for navigating the visual and/or non-visual media (see also Anselin, this volume). Being able to intelligently respond to the environment requires *analysis*, which may generate other visual or non-visual media, but the capability to alter the environment in some way requires *design* capability, which is central to many virtual worlds. At one level, all three of these processes are involved in the simple act of navigation, say, but in our context, we consider exploration to be prior to analysis and analysis prior to design, notwithstanding the fact that each of these is relevant in achieving any of the others.

Finally, we come to *connection*. Single users connected through headsets to their virtual worlds represent the extreme of direct *immersion* although single users without such connection technology but in the standalone desktop environment will probably always represent the largest class of passive VR users. Although *semi-immersion* in CAVEs and theatres loosens the hardware connection by adding more point and click in a group context, the cutting edge of VR at present is coming from the Net, from networked VR where users are 'physically' in a standalone context but are 'virtually' in the same environment. This is *remote* VR but, as in all the other dichotomies we have posed, it is never a question of immersion or not, for virtual environments can be constructed to draw on the best elements of each. One last point before we show how our case studies cover this wide domain: we have implied that all the VR we are talking of here is based on computers and their networks, but other digital media might be involved for VR, often mixing analogue and digital. Ideas of augmented reality where TV, video and computation are embedded together exist throughout the media but here we will exclusively concentrate on the digital world, avoiding consideration of how other media might be involved.

Arraying these many dimensions against one another defines a classification within which many possible types of virtual environment exist. To illustrate this, we will show how the four exemplars or case studies that we will present fall within this classification. Let us simplify these dimensions by first examining the distinction between real and fictional applications against the three varieties of connectivity – immersive, semi-immersive and remote, and then contrasting single or multiple users of these environments against information-based, science-based and design-based applications. Our first exemplar (I) begins with 2-D representations of real geographical space and extends such map media to 3-D and other pictorial forms while generalising the single-user focus of GIS to many users who can simultaneously use the system. In Figure 8.1 we show this case study first in terms of its focus on real environments represented and modelled in semi-immersive and remote forms, and then contrast the many-user orientation of such a system to its use for information and scientific analysis. The portrayal of these classes – first as real/fictional environments versus immersive/semi-immersive/remote forms of connection, and then as single/multiple users arrayed against information/science/design applications – form the first two matrices which define these classes in Figure 8.1: we will discuss the third matrix below.

Our second exemplar (II) involves virtual urban modelling in which real geographical problems are explored in a group context in a semi-immersive environment (VR theatre) by multiple users. Their interest is in the scientific understanding of location as well as in the design of optimal locations which can only be achieved through the kind of exploratory heuristic graphics that this kind of VR enables. Our third exemplar (III) is simpler in that here we are concerned with a realistic urban world but represented as a multi-user virtual world in which remote users engage in altering that world through design. These first three examples all embrace more than one user in remote contexts based on representations of real situations. Single-user VR which originated the field is increasingly less significant, for once virtual environments are possible and once networks are in place, the participation of many users interacting and communicating simultaneously and in parallel becomes the norm.

Our last exemplar (IV) radically changes this context. Fictional urban environments in which many users interact remotely define virtual worlds for which a compete record of transactions and interactions exists. When this world is an urban spatial world, then it becomes possible to use these data to simulate that world – to stand back from the virtual environment and to attempt to understand it in conventional terms – or even in virtual terms. Thus we move the entire modelling process to a meta level. No longer are we concerned with modelling a real or fictional world virtually, but with modelling the process through which the virtual world is constructed/modelled/simulated. There is an element of infinite regress in

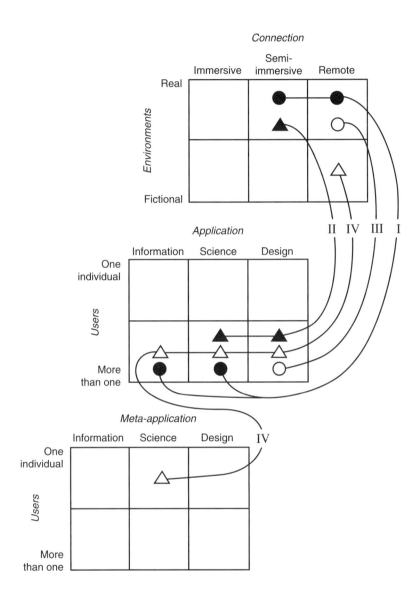

Figure 8.1 A classification of virtual environments

this recursion whereby we model digitally what is modelled digitally . . . Our final exemplar thus moves to this meta level as the third matrix in Figure 8.1 makes clear. But before we illustrate these, we need to explore the ways in which virtual environments might be represented and modelled.

8.3   Representing and Modelling Virtual Environments

The majority of virtual environments to date have been represented graphically as 3-D scenes, often rendered in considerable detail so that users are given an immediate sense of visual realism (see Curran et al, this volume, for a discussion of the realism versus accuracy of rendered versus planar Kriged surfaces). Users of such environments usually appear in visual terms as avatars, which are usually scaled to the 3-D scene and are more commonly used to represent many users in remote contexts. In immersive VR, by contrast, the emphasis is on the representation of single users in human form, and such representation is closely matched to eye, hand and other body parts.

Modelling in such contexts merges with representation. The representation of 3-D scenes is often referred to as modelling with the scenes themselves as 'models'. For example, large-scale block representations of cities are called urban or city models, not to be confused with the more generic use of the term modelling as mathematical and statistical simulation which characterises mainstream urban and geocomputational modelling. When users are introduced into such scenes, the emphasis has been on enabling them to navigate through and to manipulate objects within such scenes. These kinds of processes where real behaviour is mapped onto the virtual context is clearly 'behavioural modelling' in its widest sense, building on insights and theories from graphics, cognition, way-finding, and related areas of psychology and human-computer interaction.

However, in this context we are much less interested in these kinds of urban environment. As we have already noted, VR is rapidly moving into network and other visual and non-visual contexts where much more purposive action and decision with respect to use is being developed. Environments and users based on 3-D representations are likely to be the exception rather than the rule; VR is broadening to embrace users who do not appear formally in those environments but whose actions are outside the environment *per se* and whose behaviour patterns are much more subtle with respect to their interaction with the digital environment. A few virtual environments exist which are 2-D map-based, but their structure is not dissimilar to the norm of the 3-D environment with the exception that users usually remain passive: their location in 2-D space may be visualised but the kind of equivalence between users and environment – avatars in 3-D scenes with full motion – is not possible. In short, the traditional paradigm of VR based almost exclusively on users within scenes has little meaning once the move into other visual and non-visual environments takes place and as other forms of information and connection become important.

What is required is a broader and richer paradigm which enables the various elements of virtual environments and their use to be discussed. We do not intend to develop a fully fledged paradigm here, but we need a framework for considering the array of different representational and modelling styles that might characterise an expanded definition of VR. To this end, we consider the idea of linked information to be an important way of approaching VR. If we define VR to be the kind of environment in which users 'feel' a sense of being part of the problem with which they are concerned, in which they are able to develop a 'deeper' and more thorough understanding of such a problem through analysis and manipulation, then this casts the VR net much wider than the simple notion of working with visual environments that resemble scenes where the focus is simply upon navigation. Our augmented definition would thus include virtual laboratories, virtual lecture theatres, Web fora, and a whole host of digital simulations which would not be visual in terms of their material origins. The interface to these environments would still be largely visual but the processes of exploration and manipulation might be distinctly non-visual. The virtuality in this sense would come from representing the environment through many different kinds of information simultaneously available to the users, so linked that users are able to respond to an environment rich enough for them to gain real insights into the problems in hand.

The idea of linked information covers both visual and non-visual information. Visual information embraces multimedia in such a way that environments might be characterised by a subtle mix of live and canned photo and video, structured textual information, maps, and 3-D models of various kinds from animated scenes to fully fledged CAD models. Non-visual information embraces other senses but, in particular, textual and numerical representations and manipulations are central to such VR. The ways in which these representations might impart a sense of virtuality is likely to be through various ingenious collages and juxtapositions of the information but the concept of 'hypertext linking' or 'hotlinking' different information is central to the way these virtual environments might be constructed. In such environments, users could be actively represented – as avatars in 3-D scenes, or their presence might be portrayed in 2-D. But in general, their representation is likely to be passive, to be configured as part of the sequences of using such environments which may be recorded in some way but not necessarily visually represented within such environments.

However, the real power to interact in such environments would come from a very wide array of modelling types which might be activated by users. Four kinds of process suggest themselves: *navigation protocols* for traversing and exploring such environments in the broadest sense; *decision protocols* for interacting with other users and reaching agreement or otherwise over common problems and goals; *analytical functions* for manipulating information, using a variety of formal and informal 'scientific' procedures, many of which might be part of conventional desktop software; and *manipulation functions* which deal with ways in which information might be changed and combined and new information introduced into the environment: design, for example, would form one type of such function. We will deal with these in turn.

Navigation protocols do not deal simply with movement in 3-D space but with the entire way in which users explore the information contained within the virtual environment. For example, a user may literally move across a map, accessing 3-D scenes, photos, video clips, textual information, sounds and so on, and it is the overall process of access – the way that access is initiated, the design of that access, and the way in which users juxtapose information and generate insights – that is part of this domain: in short, this is the human–computer interface but structured with particular purposes and problems in mind. There is a clear design task here in that in modelling such environments the best ways in which such navigation leads to insights must be constructed, and this will involve much trial and errors with users, and with the representation of the environment. When many users are navigating an environment, then these protocols must extend to ways in which user paths cross, and this leads us directly to modelling protocols based on interaction and decision. Note as well that part of the process of developing such environments involves logging and charting navigation so that the history of the ways in which users have interacted with their environment can be used positively to direct further interaction.

Decision protocols involve ways in which users interact to some purpose. Like navigation, interaction can take place on different levels. Many users might be present in a virtual environment but their interaction may be passive. They may not even be aware of others as in the case, for example, of several users being logged into the same Web page. If users are aware of each other, their interaction may be entirely casual. The 'ActiveWorlds' server software illustrated below contains many virtual worlds whose sole purpose is to engage in virtual chat, where the designers have not provided any functionality above the level of 3-D and avatar represen- tation other than the ability to chat to other users who appear in the scene and who are able to articulate basic gestures such as frowning, waving, etc. These kinds of interaction can be developed in non-graphic ways, as in early versions of multi-user domains (MUDS) and IRC, but what we have in mind here is much more purposive.

The interaction that we are concerned with involves the analysis and design of urban systems and thus interaction that is important must involve extracting various abstractions from the urban scene. A good deal of this might be concerned with single-user interaction with the environment itself, but when users are required to jointly engage in some task then the interactive environment must be closely specified. There is very little research in this area as yet. Existing psychological and sociological research on non-virtual problem-solving environments is difficult to transfer to virtual environments. There is some work on collaborative spatial decision-making using GIS which heralds this kind of interaction (see Carver 1997) but there are few case studies as yet. Although many virtual environments based on multiple users are being developed at present, the emphasis is on use and appli- cation, rather than on understanding the best ways of reaching decisions. Clearly charting decisions, developing ordered structures in which information can be abstracted and interpreted, facilitating Internet conversation and so on are all issues that have to be reflected in the design of such environments.

Developing basic functions which users employ to generate insights is more straightforward as many of these already exist as part of software which is being used to construct virtual environments. Traditionally VR systems have hardly any *analytical functions* which users can employ to make sense of their environment because the emphasis, as we have seen, has been upon 3-D representation and modelling through visual navigation. However, in an analytic context, then the complete range of functionality that exists in desktop and workstation software can be adapted. Already 2-D geographical information can be distributed across the Net and environments constructed which access this (Plewe 1997). GIS software is being adapted to multi-users, and Web forum software enables users to begin to connect up to such data in a structured way. In our first case study on virtual GIS, we will show how simple desktop functionality relevant to abstracting and modelling 2-D map data is already available in a remote context, while adding analytic functionality to CAD in a Web context is under way. In fact, the development of 3-D analytic functionality based on Web software such as VRML (Virtual Reality Modelling Language), the development of platform-independent applet software and such like is well advanced. During the next five years, there is bound to be an explosion of applications stretching the concept of virtuality along the lines we have speculated here.

Our final set of modelling tools involve *manipulation functions*. Manipulation in traditional VR involves reconstruction of the 3-D scene in various ways by users from within the scene but the kinds of manipulation we have in mind here are much broader. For example, if the environment were based on some mathematical model, then manipulation might involve sensitivity testing, or control/optimisation of the model. There might be some visualisation of the solution or phase space which is tied to manipulation as we will show in the virtual urban model, the second of our exemplars illustrated below. There are many tricks which we can use to visualise such manipulations but the key issue is that considerable functionality is likely to reside behind any visualisation. In traditional VR, this has rarely been the case but, for example, even in manipulating the components of 3-D scenes, there is a need for evaluation functions which display how the manipulation meets prior goals. In turn, these must be specified and appropriately represented within the virtual environment. In short, manipulation of any kind must be to some purpose and therefore is likely to involve the use of functions which indicate how this purpose is being achieved. Finally, all four of these modelling functions – navigation, decision, analysis and manipulation – cannot be easily separated from one another for to implement one is only possible through the others. In the four exemplars that we now discuss, we will indicate the essential nature of this interdependence.

8.4   Exemplar I: From Internet GIS to Virtual GIS

We are constructing an Internet GIS for Greater London which is being linked to a variety of other visual and non-visual information. In Figure 8.1, we classify this environment as existing in semi-immersive *and* remote form, accessible to many users for purposes of information and scientific analysis. The environment can be accessed in a VR theatre on the full screen with many users able to interact with the software in a group decision-making context but with other users logged onto the software remotely. At present, the heart of the system is the distributed GIS from ESRI called ArcView Internet Map Server (AIMS). This software exists on a Web server and whenever a client (remote user) delivers a request to the server, the server fires up ArcView which processes the request and delivers the output back to the client as a Java applet called MapCafe. MapCafe has the look and feel of the desktop GIS called ArcView; AIMS can be tailored to deliver considerable functionality characteristic of conventional GIS to remote users. For example, layers can be switched on and off, zoom and pan features are standard, query functions are central to the software, while several other functions involving 3-D visualisation, network analysis and spatial analysis can be invoked.

Although the GIS is at the heart of the environment, because it is accessed through a Web browser, then all the hyperlinking associated with the Web is possible. We have exploited this through hotlinking the GIS to other visualisations of the data, namely video clips, related text and numerical information in basic tabular/page form, VRML models, and other Web pages. Because the model is in its earliest stages of construction, we have not yet begun to develop ways in which users might be represented within the environment, nor have we examined how any formal interaction between them might be structured and recorded and protocols devised for reaching decisions. The ultimate idea of the system is that information will be available for several different types of user who will have different sets of privileges enabling them to perform analyses, store and represent their own results within the system, browse the system, converse with other users for specific purposes, download data for their own use, upload data into the system, and move towards the design of planning and environmental policies at different scales. The system will ultimately link scientists, policy-makers and specific interest groups as well as being available for more general public access.

The system only exists as a demonstration within the VR Theatre context as yet but, to give some idea of its potential, we will illustrate its current working. The server which runs the Internet GIS is mounted on a highly specified PC which also contains the Web browser software through which it is accessed. The VR theatre where the system is viewable in its complete form is driven by a Silicon Graphics Reality Engine (Onyx 2) with two-channel output which generates a double-size screen with dimensions in the ratio of 12:5, the scale of the entire end wall of the theatre. The Netscape viewer is configured accordingly. The interface is in two sections. The output of the GIS as MapCafe is presented on the left hand side of the screen, always in 2-D map form; on the right hand side of the screen, various information is loaded relating first to the map and accessible as hotlinks from MapCafe – VRML models, videoclips, photos, text pages and other Web pages. However, from these outputs which are accessible only from the map, further information is then displayed in the same area of the screen from hotlinks associated with this other non-map information. So, for example, the user can hotlink from the map to a VRML model but then from the VRML model the user can hotlink to videoclips, other Web sites and so on. The map always remains on the left of the screen but the information on the right may eventually be many levels removed from the map as the user searches other information within the system as well as the Web. The navigation is thus fairly basic with the map always as the starting point and with the user always moving back to the map to initiate any new search through the trees of information accessible within the browser.

To give an idea of what is possible, we show the scale of the browser in Plate III. The left window is MapCafe which resembles ArcView but is in effect the Java applet which AIMS delivers in response to a query from the client. The layers in MapCafe (seen in legend on the left of the map) can be switched on and off in usual desktop GIS manner and, as the user zooms, different layers of information can be made active. In the scene shown, which is Parliament Square, the base map is Ordnance Survey Landline data which is laid over the Cities Revealed ortho-photographic database for London. In that scene, it is possible to load a VRML model for the entire area of Central London by clicking a hotlink in the corner of the Square. This model is loaded into the Cosmo Player which is shown on the right of the screen. It can be rotated and viewed from any angle; that shown is looking into Trafalgar Square with the Thames and South Bank in the distance. This model has been crudely developed using the 3D Analyst extension to ArcView and as yet has not been rendered to the level of realism that might be expected. It is possible for the user to enable AIMS to generate any VRML view from the OS Landline data from within the Internet GIS because this is now standard functionality in ArcView, subject, of course, to appropriate licensing.

When a user first logs on, the extent of the London data is first displayed. Figure 8.2 shows the left hand screen map which appears on entry. This is a map of the London boroughs with different sizes of town centre shown as proportional circles. In the demonstrator that we have constructed to date, it is our intention simply to illustrate possibilities and thus we have incorporated diverse data of this kind. Also shown on this screen in the centre of London are two boxes; the first defines the area where the user can zoom into the Bloomsbury district where University College London is situated and explore a VRML version of the Quad, load related video scenes and access various information about the College and the area; the second enables a zoom into central London (Westminster) and access to the scenes shown in Plate III.

To illustrate the functionality and diversity of the system, we have layers of information in the system that pertain to health care from the Riverside Health Trust which covers the Hammersmith, Kensington and Ealing areas of London. When we zoom into these areas, MapCafe presents medium scale (Bartholomew) road and rail network data and onto this we can switch relevant layers. A contour surface whose density is proportional to the number of visits by health care workers to patients is shown in the zoom in Figure 8.2, and the spider lines show how we can reallocate health care workers from their base stations to patients in such a way that travel cost is minimised. This has been done within ArcView using the standard Network Analyst software. It is illustrative only of the kind of data that can be preprocessed and displayed, but this implies the range of functions that ultimately we will enable for many remote users within the Internet software.

So far, the navigation protocols for this system are elementary. We have not yet embedded the product into a Web forum and thus the decision protocols are barely developed. However, use of the product in the VR Theatre context has been extremely successful, simply in fashioning the demonstrator and in engaging debate

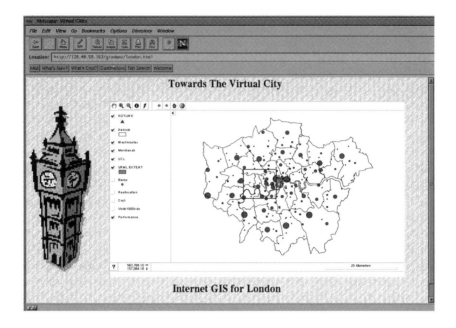

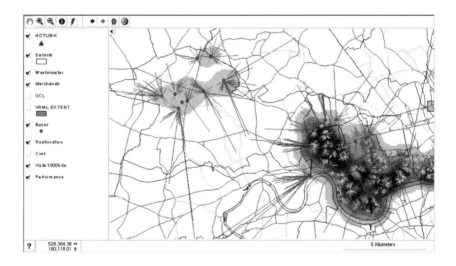

Figure 8.2 *Zooming into and displaying different layers using MapCafe*

about the whole idea. The functions so far are all those which are possible within Internet GIS and as yet there is no capability for users to produce policies or designs. However, in the various exemplars which follow, the kinds of design capability that we will illustrate there, can and may be used as this virtual GIS is further developed.

8.5 Exemplar II: Virtual Urban Modelling

Our second exemplar changes the focus substantially and is concerned with a virtual environment within which mathematical models might be better understood and solutions to problems involving their optimisation reached through visualisation involving many experts. There is a long tradition of spatial interaction modelling which is one of the cornerstones of geocomputation. Gravitational models have been used since the 1950s for all kinds of locational and transportation planning and considerable effort has been put into an understanding of their mathematical structure. As part of this, these models can be viewed as processes in which individuals optimise their location. General frameworks have been developed which show that the resultant patterns observed in cities are consistent with accessibility, entropy or utility maximising where the objective functions of access, entropy or utility are conceived as spatial averages describing the dispersion of tastes within any population. This dispersion effectively means that everyone does not locate in the same place and that the kind of distance decay patterns that characterise the way individuals interact and travel in real life are effectively replicated (Wilson et al 1981).

The fact that such models can be easily placed within a framework of optimisation means that there is the chance that optimal solutions under different constraints might be found for a variety of practical location problems in which equity and efficiency considerations can be traded off against one another. However, these kinds of model are intrinsically nonlinear; they can be solved exactly in their pure form where there is a global objective function of the kind used in their derivation, but once more realistic and practical constraints are introduced, such models become impossible to optimise directly. For example, consider the problem of a retailer or developer wishing to locate a shopping centre so that his/her profits are maximised. This problem has been handled in the past by developing a spatial interaction model which links potential shoppers at their home base to shopping centres, and exploring how much trade can be captured when new shopping centre locations are introduced into the retail landscape. This problem is usually solved by trial and error: plugging in a shopping centre and evaluating how profitable it is relative to other locations. Because there are often only a limited number of locations for new centres, then each can be evaluated and their profitability assessed.

There are versions of this problem that can be set up as formal optimisation but, in general, even with relatively simple problems such as these, it is not possible to solve the optimisation problem directly by using the model to mathematically derive the most profitable location for a centre of a given size. Real problems, of course, are much more complex. Retailers and developers face an environment in which the decision is not just to find the best location for a new store. They might expand, contract or close existing stores; they might develop different sizes of store for different market niches; they are faced with other competitors attempting the same, with a variety of planning constraints which affect what they might do, all against the vicissitudes of the market place and capital finance. They are also faced with the fact that what they might do in the short term influences the medium term. Locating a new store might attract others to it and this kind of externality affects what is profitable over different terms. In short, such problems are highly nonlinear, full of positive feedbacks in space and time which make their solution mathematically intractable. The only way forward is through the use of heuristics to get some sense of the solution space in which such a problem exists. Virtual reality systems offer much promise as we will now illustrate.

Imagine that the quest is for some developer to maximise retail turnover by locating one or several stores within an existing retail landscape. The model which drives this problem is not one with simply a spatial dimension but it has a temporal dimension in that once a store is increased in size or a new store located, then there are positive feedbacks in that the store attracts others to its location or it begins to fail. The time taken to develop the store is also critical, so the problem is one that must be solved both spatially – in terms of location – and temporally – in terms of the phasing of development. This problem resembles those formulated by Clarke and Wilson (1983). There is an immediate visualisation of the geographical retail turnover surface which will clearly change as centres change in size and location. However, when a developer examines this surface, then the location where there is maximum turnover is no longer necessarily maximum once a centre of a given size is located there. There are competition effects as well as positive feedback effects which change the surface. Add to that the kinds of constraints that characterise real conditions such as the fact that locations may be places of maximum turnover but with no space to develop or the land is owned by those who will not sell or the cost of land is too high, then these must also be built in as constraints on the optimisation. These are often hard to describe in the kind of mathematics that optimisation requires for once highly nonlinear problems are subject to very mixed constraints – integer, binary and so on – then their chance of solution is almost non-existent.

Visual solutions or rather visual solution sequences in which users can move towards better solutions are the only way forward. We are developing such an approach in a semi-immersive context. In Plate IV, we show a map of the UK town of Wolverhampton where the central area (within the ring road) has been set up as a 3-D VRML model (generated from ArcView). Draped over the entire map is the existing retail turnover surface (also generated using Spatial Analyst within ArcView) from data on retail turnover at unit postcode level (~100 m resolution), made available for a related project by the Office for National Statistics (DETR 1998). This visualisation uses Cosmo Player to view a VRML model within a Web browser and, when displayed in the semi-immersive context of the VR Theatre, it becomes immediately possible for groups of interested experts to examine the meaning of the surface and changes to it. The following kind of visualisation and optimisation thus becomes possible. The turnover surface is generated by a highly nonlinear retail model calibrated to the existing situation. The inputs to this model are size and location of centre, travel network and the density of consumers (population), all of which can be visualised in the manner of Plate IV.

As these inputs are changed, the surface will change and thus the environment we have under construction is one in which the inputs to this model can be manipulated directly by users who can then view the outputs of the model directly on the screen. Size of centre can be indicated in the 3-D block model, density of population we can assume is fixed, but the route network can also be manipulated graphically. Assuming that only the size and location of centre or store were to change (the developer's location problem), then it is possible for the user to drag a centre to different positions in the map and, as this process takes place, the model is computed on-the-fly and the surface redrawn. At SG Reality Engine refresh rates (computation at 10 million polygons per second), the surface changes at speeds in excess of the time it takes to move the mouse and thus real-time motion is engendered. The user can also change the size of the centre using various sliders and the program is so configured to stop the user infringing any basic constraints where location is forbidden or not feasible.

Optimisation is achieved visually. As the location and size of the centre changes, so does the retail turnover surface and what might look like a good location at one instant may not be once the centre is dragged to it. To make progress in moving to a better location, then a variety of other information must be displayed and computed. For example, the derivative surfaces can be displayed, showing best directions of search and the entire armoury in visualisation of phase spaces employed to illustrate progress. Of course, this visualisation has assumed that the objective function is retail turnover *per se* but there may be other related measures which cannot be visualised geographically. In fact, this problem needs to be researched as a problem in scientific visualisation as well as one in virtual optimisation, and, to this end, it is likely that the ultimate interface will not only consist of a geographical phase space as in Plate IV but also more abstract windows on the model as well as sliders and other devices for controlling the optimisation. We are exploring a variety of VR software appropriate for this implementation.

Finally, we should note some features of the modelling styles used here. There is extensive functionality that users would make use of here including many GIS-like functions based on statistical and modelling procedures; these would be linked to ways of manipulating the model visually, mathematically, and through its optimisation. The 2-D and 3-D representation is fairly standard and there are well-developed techniques as we show in Plate IV for these, but the decision and navigation protocols are not really specified. This environment so far relies on the expertise of users seated in the material world – in the VR Theatre or in whatever environment this type of software is displayed – and there is clearly considerable work to be done of the way users begin to respond to the optimisation process. This is another illustration of the fact that virtual environments are so new that, until we begin to use them, we will have little idea as to how to design them in the most effective way.

8.6   Exemplar III: Virtual Urban Design

The most well-developed virtual environments in the visual sense are those in which both environment and users are visually configured with users remote from one another materially, but together virtually. These are those virtual worlds which are sprouting up everywhere on the Net (Dodge and Smith 1998). We will illustrate what is possible from AlphaWorld, one the largest worlds (circa 200 000 users) on the ActiveWorlds server, but before we do this, we should set the context for urban design. One of us (Andy Smith) has developed a 2-D to 3-D virtual urban information system which enables users to tour selected sites in central London by navigating from points on a map through 3-D scenes captured as wrap-around panoramas from which access to other relevant Web sites can be made. The software is called Wired Whitehall and one version hangs off our Webpage at *http://www.casa.ucl.ac.uk/* (see Batty 1997a). The idea of hotlinking map sites to 3-D scenes is a basic tool which we are using in our extension of Internet GIS to virtual GIS which we presented above, but Wired Whitehall is now being further developed using techniques and ideas which are used in virtual worlds software.

There are three basic components of a virtual world: a 3-D environment which is represented in VRML-like form but which can also be manipulated by users; representation of users who log on remotely as avatars who are able to interact with each other through primitive visual gestures; and windows in which users (as avatars) can converse through text appearing as script which is a record of all that passes 'conversationally' within the world. Such a world is illustrated in Plate V. The chat box and the window which list those within the world are clearly visible below the 3-D representation. Three avatars are shown in the scene and the record of conversation is typical of casual interaction within such worlds. But the ability to manipulate the 3-D scene is also shown. A user can only manipulate or design on a plot which he or she owns and this involves being a fully paid up member of such a world. Basic manipulation of 3-D shapes is very easy: on AlphaWorld, Andy Smith's Online Planning Office was and is under continual construction, except during the period when his subscription ran out and the plot was immediately 'destroyed' by the owners of the world!

Meshing this kind of environment with the kind of interface in Wired Whitehall involves making the 3-D representation in the virtual world considerably more realistic. It is possible to combine digital photo and video with objects which can be moved within scenes. A good example of this is shown in Plate VI where a panorama of residential development around the Surrey Docks, activated using the same kinds of software used to enable its animation on the Web (as in Wired Whitehall), is displayed together with a 'bouncing ball'. The ball which bounces all the time can be positioned anywhere within the scene by the user and this contains all the elements necessary for certain types of detailed urban design. Andy Smith is developing the Wired Whitehall virtual tourist into a virtual world in which any user who logs on and navigates into the same scene as any other user, will appear as an avatar, can engage in 'chat', and can move certain objects within the scene, even if these objects are not under their individual ownership. This is the basis of virtual urban design. If the chat is purposive, to locate features within the urban scene which represent the best designs that those in the scene can develop, then design can take place virtually. An illustration of what we are moving towards is shown in Plate VII where the object of interest is a telephone box to be located somewhere within the 360 degree panorama at the junction of Whitehall with Whitehall Place. A combination of cursor keys, mouse and menu items enables the user to control his or her own motion in the scene as well as the motion and location of the telephone box.

There are several features of this interface that require comment. The navigation protocol is straightforward and is that adopted as standard in virtual worlds software. This is not something that we intend to change very much but the placement of objects within the scene to optimise design is a feature that we are developing. However, because these manipulation functions involve design, we are considering how we might develop some more abstract characterisation of what constitutes good design in such a world. At present, the evaluation is entirely subjective, based on agreement. Moreover, there is no necessary basis for agreement other than the fact that presumably two or more users must agree that there is a design problem to be solved. We may decide to incorporate some stronger rules into the interface to achieve some convergence but all this implies that there should be different pathways to decision that users can invoke, one of which must always be 'no' agreement. As yet, there are no analytic functions which users can generate in their quest to design but we are considering adding features which analyse scenes from a qualitative viewpoint in terms of perspective views and such like. Text-based information can, of course, be used to supplement such visual interfaces. But we see the greatest progress being made by synthesising real video and photo with VRML models through which avatars can move and from which we can extract and display to users basic information about the environment's geometry.

8.7 Exemplar IV: Modelling the Morphology of a Virtual World

Our last example is very different. The three exemplars so far all illustrate how users can begin to generate insights in geocomputational terms about the urban world by entering those worlds, engaging in dialogue, extracting information, and manipulating the worlds through dialogue and intelligence derived from information. All three of our virtual environments have these features to a greater or lesser degree. However, there are a set of problems of understanding virtual environments that exist at a meta level. For example, when we enquire into the best ways of designing protocols and functions, then these types of human–computer interface problem are at one level removed. Yet most of these meta problems are not urban ones *per se* although there is one meta problem which spins off from these kinds of environment which does involve urban modelling. Many virtual worlds are urban worlds in cyberspace, but sufficiently like the real urban world to be subjects of understanding in their own right. Our three exemplars do not fall into this class; they are already abstractions of the real world into the digital and there is little purpose in thinking of these as urban worlds to be studied in their own right for much of what characterises these environments is designed to elicit insights into the real world. But urban worlds which are entirely fictional are worthy of study (Macmillan 1996). Our last exemplar introduces these.

AlphaWorld, as we have noted, is the biggest virtual world we know of, certainly the biggest running on the ActiveWorlds server. It is essentially an urban world where the emphasis is on the ownership of land and where a city structure has clearly grown up. There is no real distance in AlphaWorld in that it is possible to teleport anywhere within this world, and thus one of the most distinctive structuring factors of real urban space is missing. However, AlphaWorld appears to have grown as real cities have grown. Its morphology as computed in December 1996 is shown in Plate VIII, where it is clear that it has a distinct CBD (Central Business District) (ground zero), and distinct axes in the eight compass quadrants where development has clustered. From the ActiveWorlds homepage (*http:// www.activeworlds.com/events/satellite.html*), the owners of the world say:

> You can see the 'starfish' shape of the building as people crowd their buildings along the north–south axis and 'equator' of AW, and as they build along the coordinates with matching numbers. (i.e. 220n, 200w, 450s, 450e etc.) Some do this so that their coordinates are easy to remember, and others simply by building onto what others have built.

This is a fascinating quote in that clearly these factors – imageability and agglomeration economies – are important, but perhaps AlphaWorld reflects the quest for space too – for isolation as well as congestion, the two forces which define the tension of the modern city. The challenge is thus: can morphological models based on simple processes of attraction and repulsion, concentration and dispersion, be built which reflect the way AlphaWorld has developed? And can the processes of decision which define this growth and change be linked to such models, thus establishing the relative importance of the various urban forces and their relationship to different populations types? Are there several processes at work and is there spatial differentiation of these within the morphology which has developed?

What is remarkable about these virtual worlds is that the morphology mirrors in some sense real cities but, unlike real cities, there is a complete record of what has taken place in these worlds from the server log, if it is not destroyed. This provides a potentially rich source for not only mapping and understanding urban cyberspace but also for beginning to speculate on whether the kinds of processes that take place in fictional worlds throw insight on processes taking place in the real world. In modelling real urban morphologies, distance is all important and the entire heritage of urban theory from central place theory and spatial interaction to fractal geometry is built around the notion that location can be explained as a trade-off between mass (size) and distance (travel cost). In operational models that have been built, one of the features that has been examined and has always proved curious is that if the distance effect is relaxed entirely, and thus is no longer relevant, then these models still generate morphologies that are close to those observed. In short, the question as to how important is distance in locational decision-making in contrast to the notion that cities grow out from the core by simply adding buildings on available land closest to the growing mass, is still largely undecided. AlphaWorld and its like could well provide insights into how such forces play themselves out.

It is unlikely that study of AlphaWorld *per se* would yield much insight into these questions unless a very detailed log of all decisions were available. It is much more fruitful to consider setting up a new urban world for purposes of study where the decision-making process is known in advance and where distance may be a factor in location. This would ensure that the inventors or owners of such a world would structure the log of data in such a way as to be amenable to urban analysis and simulation. The problem would be in planning such a world to take off in the unplanned fashion that has led to the kinds of organic growth characteristic of AlphaWorld. This could happen but it also suggests that many such worlds should be examined, for doubtless it will be possible to classify them into different types. The kinds of model that could be used to simulate these worlds are those based on local action such as in cellular automata (CA: see Burrough, this volume). CA are able to simulate growth and diffusion under very different sets of rules which could be adapted easily to the rules that are present in worlds like AlphaWorld. For example, the morphology of AlphaWorld looks like crystal growth around a core or seed, and there is a class of such models which are able to generate similar structures where the forces which determine where unoccupied cells or vacant land 'crystallises' into building can be tuned to reflect the trade-off of agglomeration and dispersion. In fact, CA models ignore global distance effects, generating growth in such a way that global pattern is always a result of local action (Batty 1997b).

Notwithstanding the design of the world and its placement on a server which can log all interactions, the models we might use to explore this world exist themselves in a digital environment. It is quite possible that the model of the virtual world is itself developed as a virtual environment, in the manner of the retail location model (Exemplar II). This poses an interesting regress – to use VR to understand VR – and, of course, one might imagine the regress being pursued to several orders, or more likely different aspects of the world and its model might be explored to different orders of regress. At present, we have the design of such an urban world under scrutiny with the obvious possibility that all kinds of research into this urban cyberspace might result, from its conventional urban geography, to fully fledged mathematical simulation, and even to its cyber-politics.

8.8   Next Steps

If we examine the classification we originally introduced in Figure 8.1, there are areas where we have no experience of virtual environments but which appear relevant to the wider study of urban applications. In this chapter, we have largely ignored immersive worlds and worlds which are populated by single users, instead concentrating on environments which are much more open to many users linked together in the flesh or over the Net or both. Furthermore, we have only barely touched on fictional worlds which have enormous potential for the development of theory. Theory in the form of hypothetical models always contains elements of fiction, and this is often the domain of single individuals. Virtual laboratories which bring resources to theory development in a remote context across the Net, or virtual field courses through which individuals learn about data, their interpretation, and their simulation, are developments which we have not sketched here but which must clearly be considered as examples of virtual human environments.

In terms of the exemplars illustrated here, each one is under continual development. The entire area is under very rapid development at present and many of the ideas, particularly those of linking different media and of displaying this media remotely across the Web, have only become possible in the last few months. In terms of our four modelling themes based on navigation, decision, analysis and manipulation, then so far there has been very little work on the kinds of navigation and decision which multiple users require when interacting within such environments. This requires very explicit and targeted research, somewhat different from the kind of applied research which is necessary to construct such environments. Throughout computing, such research has always been low key and somewhat small-scale. The single-user interface which has dominated computation to date has always been left in the hands of the users and rarely do software designers tailor their designs to questions of human performance and problem-solving behaviour. Now that computation is beginning to involve multiple users, such research into the human–computer interface takes on a new urgency, and it is through the kinds of virtual urban environments introduced in this paper that progress in this domain will be made.

References

Batty M 1997a Digital planning: preparing for a fully wired world. In Sikdar P K, Dhingra S L, Krishna Rao K V (eds) *Computers in urban planning and urban management: keynote papers.* Delhi, India, Narosa Publishing House: 13–30

Batty, M 1997b *Growing cities.* Unpublished paper, University College London, Centre for Advanced Spatial Analysis

Carver S 1997 Using (geo)graphical environments on the World Wide Web to improve participation in social science research. In Mumford A (ed.) *Graphics, visualisation, and the social sciences.* Loughborough, Technical Report 33, Advisory Group on Computer Graphics: 11–14

Clarke M, Wilson A G 1983 Dynamics of urban spatial structure: progress and problems. *Journal of Regional Science* 21: 1–18

Cruz-Neira C, Sandin D J, DeFanti T A 1993 Surround-screen projection-based virtual reality: the design and implementation of a CAVE. *Computer Graphics (SIGGRAPH) Proceedings*, Annual Conference Series: 135–42

DETR 1998 *Town centres: defining boundaries for statistical monitoring.* London, Department of Environment, Transport and the Regions, HMSO

Dodge M, Smith, A 1998 *Virtual Internet design arenas: the potential of virtual worlds for urban design teaching.* Unpublished paper, University College London, Centre for Advanced Spatial Analysis

Heim M 1997 The art of virtual reality. In Droege P (ed.) *Intelligent environments: spatial aspects of the information revolution.* Amsterdam, North Holland Publishing Company: 421–37

Kalawsky R 1993 *The science of virtual reality and virtual environments.* Wokingham, Addison-Wesley

Kitchen R 1998 *Cyberspace: the world in wires.* Chichester, John Wiley

Macmillan B 1996 Fun and games: serious toys for city modelling in a GIS environment. In Longley P, Batty M (eds) *Spatial analysis: modelling in a GIS environment.* Cambridge, GeoInformation International: 153–65

Negroponte N 1995 *Being digital.* New York, Knopf

Plewe B 1997 *GIS online: information, retrieval, mapping, and the Internet.* Santa Fe, OnWord Press

Rheingold H 1991 *Virtual reality.* New York, Touchstone Books

Webber M M 1964 The urban place and nonplace urban realm. In Webber M M (ed.) *Explorations into urban structure.* Philadelphia, University of Pennsylvania Press: 79–153

Wilson A G, Coelho J D, Macgill S M, Williams H C W L 1981 *Optimization in locational and transport analysis.* Chichester, John Wiley

Part Five
SPACE–TIME DYNAMICS

9

Dynamic Modelling and Geocomputation

Peter A Burrough

Summary

This chapter begins with a discussion of the emergence of distributed modelling (specifically finite-element, finite-difference and cellular automata models) for computing the transport and distribution of materials over space. Next it provides a review of some computational methods for distributed modelling, and the ways in which physical process models have been coupled to GIS. Trends in computational methods are seen as matching, or even outstripping, improvements in modelling skills, yet together these developments are making desk-top dynamic modelling systems for all a reality. The principal problems for the future centre upon basic questions about the morphodynamics of spatial change. These in turn raise issues of errors and uncertainties in data and model parameters, of model resolution, and of the representation of chaotic, unstable behaviour in nonlinear systems.

9.1 Introduction – GIS Technology for Spatial Analysis and Dynamic Modelling

Today, geographical information systems (GIS) form a mature technology for storing, organising, retrieving and modifying information on the spatial distribution of plants, natural resources, forest areas, land use, land parcels, utilities and many other natural and anthropogenic features (Burrough and McDonnell 1998). The challenge for the coming years is to transform these essentially static systems into tools that are capable of providing realistic and affordable insights into space–time processes, both in scientific research and resource management. Although most GIS

Geocomputation: A Primer. Edited by Paul A Longley, Sue M Brooks, Rachael McDonnell and Bill Macmillan.
© 1998 John Wiley & Sons Ltd.

can retrieve, display, count, accumulate, cut, split, stick together and transform spatial data, many commercial systems do not support computational modelling of the kinds of dynamic processes encountered in the Earth and biological sciences, such as hydrology or ecology. Instead, GIS are often used to supply data to, and display results from stand-alone models that run separately from the GIS. Computationally this operation may involve little more than ensuring a suitable data interface, but for the scientist there are several reasons why this procedure may be unsatisfactory.

Several difficulties arise when linking dynamic models to a standard GIS. First, there may be important conceptual differences between the ways modellers, data collectors and GIS technical persons perceive the world and how their view of reality should be structured and organised in the computer (Burrough 1996). Second, the data in the GIS may not be recorded or stored in the most suitable form for the model and may need to be converted – e.g. the data are coded in polygons when information about continuous spatial variation is required. Third, the model may be 'hard-wired', which means that the researcher cannot modify the procedures and algorithms should these be inadequate or inefficient. Fourth, unless one is skilled in computer programming, it is difficult and time consuming to write dynamic models of spatial processes quickly and efficiently. Fifth, most visualisation methods in GIS do not support interactive space–time presentation of model results. Finally, methods of open systems modelling, or of cellular automata, which meet many ecologists' requirements to model dynamic processes quickly and efficiently, are rarely implemented in GIS (Clarke and Olsen 1996; Park and Wagner 1997; Takeyama and Couclelis 1997).

In this chapter I review briefly the basic principles of modelling spatial dynamic processes and then describe how these principles have been embodied in a high-level computer programming language called PCRaster (Deursen and Wesseling 1995). Demonstration examples of the diffusion of plants over a landscape and the flow of surface water during storms in small catchments illustrate the language and demonstrate its power and simplicity. No system of modelling is a perfect representation of reality, however, and an important issue in many environmental applications is the effect of uncertainties in the data on the results. This issue is discussed in terms of artefacts that may arise through the derivation of topological flow nets from digital elevation models (DEMs). Finally, it is shown that some dynamic process cannot be modelled in a satisfactory way without uncertainty – to model them deterministically is not to understand the process correctly.

9.2   The Basic Principles of Dynamic Spatial Modelling

A dynamic spatial model may be defined as a mathematical representation of a real-world process in which the state of a location on the Earth's surface changes in response to variations in the driving forces. Any system for modelling space–time processes must include procedures for discretising space–time, and for the computation of new attributes for the spatial and temporal units in response to the driving forces.

9.2.1   The Discretisation of Space–Time

There are basically two contrasting and complementary ways to discretise space, namely to recognise objects in space, or to divide space into a series of tiles. Conventionally 'objects in space' are described by vector data entities such as points, lines and polygons, while continuous fields may be described in terms of networks linking spot heights (e.g. the TIN model) or by regular tessellation (Burrough and McDonnell 1998). Of the latter, regular grids (rasters) are the easiest to compute. They are used in this chapter, though other scientists prefer to use spatial units (finite elements) of irregular form that match the changes in form of the area being modelled (e.g. Bates and Anderson 1993). Computationally, a gridded continuous field can be regarded as being made up of simple entities that are spatially isomorphic and which can easily be coded in a database.

Like space, time can be divided in different ways. Computationally the easiest way is to discretise time into equal steps, and that is the procedure followed in this chapter.

When discretising both time and space, the choice of the size of the interval or cell may be extremely important because variations that occur within the dimensions of the cell will not be registered by either the data or the process. Therefore it is essential to choose the correct spatial and temporal resolution. Vector systems may be better than grids at dealing with variable levels of resolution at the cost of more complicated programming. In this chapter I restrict discussion to dynamic modelling with regular gridded data.

9.2.2   The Basic Classes of Mathematical Operations for Dynamic Spatial Modelling on Grids

Computationally, dynamic modelling involves computing a new *state* (attribute value) of an entity in response to information from a time series or from a driving force included in a mathematical model. A sensation of true movement can be obtained by successively displaying each state rapidly on a computer screen.

Each entity ('object' or grid cell) is described by three kinds of information, namely *what is it?*, *where is it?* and *what is its relation to other entities?* The nature of an entity is given by its *attributes*, its whereabouts by its geographical *location or coordinates* and the spatial relations between different entities in terms of *proximity* and *connectivity* (topology).

New attributes can be computed for each grid cell from attributes at the same location using any standard mathematical operation (Burrough and McDonnell 1998). These entity related computations are termed 'point operations' because no spatial interactions are involved.

Neighbourhood interactions such as:

- interpolation;
- spatial filtering;
- derivation of first and higher order derivatives;

- derivation of surface topology: drainage networks and catchment delineation;
- contiguity assessment (clumping);
- nonlinear dilation (spreading with friction); and
- computation of viewsheds, shaded relief, and direct incident solar radiation;

provide means of computing new cell attributes as a simple function of surroundings. These operations are described in detail by Burrough and McDonnell (1998) and are to be found in many standard GIS. The difference with a dynamic GIS is that these operations can be linked together to provide a logical sequence of commands that is driven by data from time series of observations at points or over areas.

9.2.3   Computational Aspects of Dynamic Models

In dynamic spatial models we distinguish between time-driven processes that operate only at a point location and processes that operate over 2-D or 3-D space.

9.2.3.1   *Point Processes without a Memory*

With simple *point processes,* the output or state of a cell S for attribute i is determined by operations on a time-dependent variable j that provides input to the cell (Figure 9.1a).

$$S_i(t) = f(I_j(t)) \tag{1}$$

where $f(I_j(t))$ is some mathematical function operating on I_j at time t.

Deterministic or stochastic inputs   The inputs I_j to the cell may be deterministic or stochastic variables. If they are deterministic, then the values of I_j are the result of a regular, clockwork-like process. If this is so, the time series of I_j can be replaced by a mathematical function, such as a sine curve, for example. The inputs to equation (1) are then made up of the *parameters* of the sine function, which are its initial value (or phase), its wavelength and its amplitude. Once the parameters of the function have been set, all values of S_i are completely defined for all t.

Alternatively, the inputs may be a random series of numbers. The definition of a random series is one that cannot be represented by any known function – the series is itself the most compact representation of the variations observed. The driving process behind the inputs has no known physical explanation.

In many situations we cannot say whether the observed variation is deterministic or random. The variation may be too complicated to capture entirely with a mathematical function, yet it may display certain degrees of regularity or trends that can be modelled deterministically. Empirical models, usually based on linear regression of experimental data, attempt to 'explain' observed variation as a combination of deterministic components or 'trends' and unexplained 'noise'. The

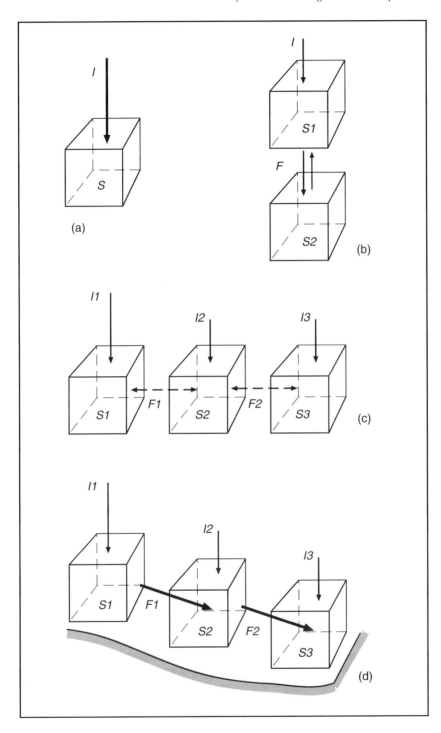

Figure 9.1   *Interactions between inputs (I), storage (S) and fluxes (F) in simple systems*

trend functions are characterised in terms of parameters such as the intercept and the weighting factors given to each contributing variable.

9.2.3.2   Point Processes with a Memory

In the above, the state of the cell is determined solely by the inputs. Now consider the situation in which the state of the cell S_i at time $t+1$ retains some information from its state at time t; the cell is said to have a memory or 'feedback' as well as the input I_j. The 'memory' of the cell may be determined by a function g operating on the initial state:

$$S_i(t+1) = g(S_i(t)) + f(I_j) \tag{2}$$

The memory may also be described in deterministic or stochastic terms. The simplest form of deterministic memory is where the function g accumulates the values of the inputs over the time series. Nonlinear, deterministic feedback models may be mathematically unstable, giving rise to mathematical chaos that may be difficult to distinguish from random noise (Stewart 1989).

Driving the state of a cell with random inputs that accumulate over time with

$$S_i(t+1) = S_i(t) + \varepsilon \tag{3}$$

where ε is a normally distributed (Gaussian) variable with zero mean and unit variance produces the Brownian trail for which correlations in values may be derived over any period of time. Equation (3) is also known as a first-order *autoregressive process*, or Markov Chain (Chatfield 1980). The strength of the memory in an autoregressive process is given by the autocovariogram, which provides a measure of the length of time that a given input value persists in the model's memory.

9.2.3.3   Vertical and Horizontal Spatial Interactions

Now consider a point process with a memory, but one in which the value of the state of the cell is limited to a maximum. A simple example is the input of rainwater on the soil. When the soil cannot absorb any more water its storage/memory is full and any excess must be redistributed. Redistribution may occur vertically by flow to greater depths (Figure 9.1b) or by lateral redistribution by diffusion or directed flow. Directed flow requires knowledge of a spatial topology or linkage between one location and another.

In the real world the state of the cell is not just determined by the inputs and the memory but by what is going on around it. Interactions with the neighbours can change the state of the cell, and the state of the cell can change the interactions with the surroundings. For gridded data it is easy to define the neighbours. In one dimension a cell has 2 neighbours, in 2-D a cell has 8 neighbours; in 3-D a cubic cell has 26 neighbours. Generally a grid cell has $(3^d - 1)$ first neighbours, where d is the dimension.

Consider the 1-D situation in Figure 9.1b in which the neighbours are arranged in a vertical column. If too much water accumulates in the top cell and its storage capacity is exceeded, gravitational flow (which can be described by Darcy's Law) causes a flux from the upper to the lower cell. The state of the lower cell is then determined by the flux (inputs from above) and its initial state. These operations are computationally easy in raster GIS.

Now consider cells that are horizontally adjacent along the transect (Figure 9.1c). If we only consider spatial adjacency as a criterion for the transfer of matter we have a dispersion process which describes the movement of a 'surplus' from one cell to the other. The rate of the dispersion may be controlled by the levels of the states of each cell, by the distance the material has to travel (a function of cell size) and by the internal resistance (or friction) for each cell. Now the state of any given cell is determined by the external inputs, the flux from the neighbours and its previous state or memory. As before, the inputs may be derived from a deterministic or stochastic process, and the redistribution of the material from cell to cell via the fluxes can also be described in deterministic or probabilistic terms.

$$S_i(t+1) = S_i(t) + F_{in}(t) + \varepsilon \qquad (4)$$

where $F_{in}(t)$ is the flux of attribute i from the n neighbours for time step t.

Directed topology   A special case of lateral or vertical transport is when the cells are connected by direct topological links that determine the direction and linkages of the fluxes (Figure 9.1d) and the random term disappears. Topological links are often explicitly built into mathematical models of flow processes (finite-element modelling) in order to represent a physically based transfer of material (usually water) under gravity over a complex surface such as a landscape (cf. Bates and Anderson 1993).

The advantage of a fully connected topology is that all movement of material from cell to cell (the flux) is controlled, subject to the law of conservation of mass. The flux can be modified by the state of any other attributes of the cells, such as surface roughness or infiltration capacity. A complication with modelling the gravity-driven flow of material, however, is that, once moving, the material acquires kinetic energy. The momentum of the moving mass may locally exceed the ability of a connected topological net and flooding of neighbouring areas may occur. Another aspect not taken into account here, and important in the transport of sediment along stream beds, is the blocking of flow by material in front of the moving mass, which may give rise to stable riffles and bars. This aspect of surface flow is handled by the concept of kinematic waves (Langbein and Leopold 1968) but is not treated further in this chapter (but see Kirkby et al 1987).

9.2.3.4   Spatial Functions Commonly Found in Raster GIS

Both diffusion and topologically directed transfer are easy to compute in 1-D, but 2-D surfaces are more difficult because of the larger number of neighbours and possible routes, and this is where GIS functionality is useful.

Dispersion   As a static process, the 2-D dispersion problem can be addressed using the well-known 'spread' algorithm and its nonlinear form, the 'spread with friction' operation, which is common in raster GIS. Computationally, all that needs to be added to make a dynamic model with these functions is to link them with time series inputs to the cells so that the usual static buffering behaves as a series of ripples.

Interpolation   Gradually varying surfaces can easily be computed by interpolation from data at point locations. The choice of methods is large and different methods may give different results (Burrough and McDonnell 1998). The variation of a surface over time can be shown by computing and storing the interpolated surface for point data from each time step, and then displaying the results as a video. Again, it is necessary to link the interpolation to the time series both for inputs and display.

Spatial filtering, windowing and 'whole surface' operations   Most raster GIS include many kinds of spatial filtering operations that operate on a square neighbourhood to compute a new attribute for the central cell. Among these are the computation of the following for the central cell of a $n \times n$ square window (Burrough and McDonnell 1998):

- mean, mode, median, diversity, minimum, maximum, range, standard deviation, or edge attributes;
- first- and second-order derivatives of a continuous mathematical surface: slope, aspect, profile convexity, plan convexity; and
- whole surface properties such as shaded relief, incident solar radiation and viewsheds.

By driving these computations with time series data one can model processes over time. For example, computing incident solar radiation for a DEM for every daylight hour yields a sequence of surfaces that can be displayed as a video of the diurnal variations of direct sunlight falling on the surface (Burrough and McDonnell 1998). None of these window operations control lateral transfer of material, however.

For DEMs in vector GIS the 2-D topology is often given explicitly by a triangular irregular network (TIN), which could be used to direct flows of material from one location to another. For altitude matrices (a digital elevation model in gridded form) the flow topology is implicit in the differences in elevation between neighbouring cells, but it needs to be extracted in order to model transport processes over the surface as envisaged in Figure 9.1d.

There are various algorithms for computing this 'local drain direction' (LDD) network. Of these, the D8 algorithm that routes flow from the cell at the centre of a 3×3 window to its steepest downhill neighbour is best known, though many authors believe that it is not the most realistic because it discretises flow directions in units of 45 degrees (Burrough and McDonnell 1998). Once created, however, the

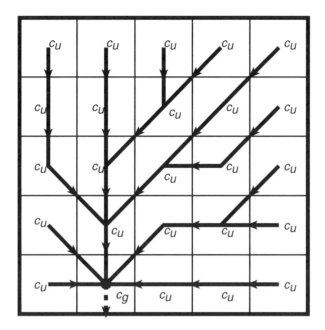

Figure 9.2 Example of a 'local drain direction' network derived from a gridded DEM

LDD net provides an approximate, discretised topology for linking cells so that the transfer of matter from one cell to the other over 2-D surfaces can easily be computed (Figure 9.2).

9.2.3.5 Moving Material over the LDD Net: Accumulation Functions

Once an explicit surface topology has been derived from a DEM and the inputs to each cell on the grid can be supplied from time series, material can be moved over the net in several ways. The most important, are the *accuflux/state*, *accucapacity-flux/state*, *accufractionflux/state*, *accuthresholdflux/state* and *accutriggerflux/state* operations. All of these compute both the flux and the state for a given cell and time loop.

First we need some definitions:

- The target cell c_g is the cell to which all flow is ultimately directed. Any cell above the target cell is called an upstream cell c_u. The upstream cells of a target cell c_g are c_u, $u = 1 \ldots n$, where n is the number of upstream cells.
- The state of any cell g at time t for an attribute i is $S(c_{gi},t)$. The flux out of a cell i for a given attribute j is $F(c_{gi},t)$.
- The *transport capacity* of a cell g for attribute i is given by TC_{gi}.
- The *transport fraction* of a cell g for attribute i is given by TF_{gi}.
- The *threshold value* for a cell g for attribute i is given by D_{gi}.
- The *trigger value* for a cell g for attribute i is given by TV_{gi}.

The *accuflux* function computes the new state of the attributes for the cell as the sum of the original cell value plus the cumulative sum of all the upstream elements draining through the cell

$$F(c_{gi}) = S(c_{gi}) + \sum_u S(c_{uj}) \tag{5}$$

If the attribute *i* for each cell is its area, the result gives the *upstream contributing area map*. When all material is transported the new state is zero, i.e. $S(c_{gi})$ at $t+1 = 0$.

The capacity, fraction, threshold and trigger operations explicitly divide the fluxes and cell states at time $t+1$ according to cell-specific parameters that control the way the flux is computed. The capacity function limits the cell-to-cell flux by a channel capacity attribute; the fraction function transports only a given proportion of material from cell to cell, the threshold function transports material only once a given threshold has been exceeded, and the trigger function transports nothing until a trigger value has been exceeded (at which point all accumulated material in the state of the cell is discharged to its downstream neighbour).

The *accucapacityflux/state* operation modifies the accumulation of flow over the network by a limiting transport capacity given in absolute values. Flow only occurs up to the level of the TC_i.

The flux per cell for attribute *i* is given by:

$$\text{If } [S(c_{gi}) + \sum_u S(c_{ui})] \leq TC_{gi} \text{ then } F(c_{gi}) = S(c_{gi}) + \sum_u S(c_{ui}) \tag{6a}$$

$$\text{If } [S(c_{gi}) + \sum_u S(c_{ui})] > TC_{gi} \text{ then } F(c_{gi}) = TC_{gi} \tag{6b}$$

The storage (new state of the cell at time $t+1$) is given by:

$$\text{If } [S(c_{gi}) + \sum_u S(c_{ui})] \leq TC_{gi} \text{ then } S(c_{gi})_{t+1} = 0 \tag{7a}$$

$$\text{If } [S(c_{gi}) + \sum_u S(c_{ui})] > TC_{gi} \text{ then } S(c_{gi})_{t+1} = [S(c_{gi}) + \sum_u S(c_{ui})] - TC_{gi} \tag{7b}$$

The *accufractionflux/state* operation limits the flow over the network by a parameter which controls the proportion of the material that can flow through each cell.

The flux per cell is given by:

$$F(c_{gi}) = [S(c_{gi}) + \sum_u S(c_{ui})] * TF_{gi} \tag{8}$$

The storage (state of the cell after transport) is given by:

$$S(c_{gi}) = [S(c_{gi}) + \sum_u S(c_{ui})] * (1 - TF_{gi}) \qquad (9)$$

The *accuthresholdflux/state* operation modifies the accumulation of flow over the network by limiting transport to values greater than a minimum threshold value per cell. No flow occurs if the D_{gi} is not exceeded.

The flux per cell is given by:

$$\text{If } [S(c_{gi}) + \sum_u S(c_{ui})] \leq D_{gi} \text{ then } F(c_{gi}) = 0 \qquad (10a)$$

$$\text{If } [S(c_{gi}) + \sum_u S(c_{ui})] > D_{gi} \text{ then } F(c_{gi}) = [S(c_{gi}) + \sum_u S(c_{uj})] - D_{gi} \qquad (10b)$$

The storage (new state of the cell at time $t + 1$) is given by:

$$\text{If } [S(c_{gi}) + \sum_u S(c_{ui})] \leq D_{gi} \text{ then } S(c_{gi})_{t+1} = [S(c_{gi}) + \sum_u S(c_{uj}) \qquad (11a)$$

$$\text{If } [S(c_{gi}) + \sum_u S(c_{ui})] > D_{gi} \text{ then } S(c_{gi})_{t+1} = D_{gi} \qquad (11b)$$

The *accutriggerflux/state* operation only allows transport (flux) to occur if a trigger value is exceeded, otherwise no transport occurs and storage accumulates.

The flux per cell is given by:

$$\text{If } [S(c_{gi}) + \sum_u S(c_{ui})] \leq TV_{gi} \text{ then } F(c_{gi}) = 0 \qquad (12a)$$

$$\text{If } [S(c_{gi}) + \sum_u S(c_{ui})] > TV_{gi} \text{ then } F(c_{gi}) = S(c_{gi}) + \sum_u S(c_{uj}) \qquad (12b)$$

The storage (new state of the cell at time $t + 1$) is given by:

$$\text{If } [S(c_{gi}) + \sum_u S(c_{ui})] \leq TV_{gi} \text{ then } S(c_{gi})_{t+1} = S(c_{gi}) + \sum_u S(c_{uj}) \qquad (13a)$$

$$\text{If } [S(c_{gi}) + \sum_u S(c_{ui})] > TV_{gi} \text{ then } S(c_{gi})_{t+1} = 0 \qquad (13b)$$

9.2.3.6   Time Aspects

All of the above can be used in time series modelling. Simple time series modelling involves repeating the function p times with p different parameter values to give p different outputs. Accumulation modelling with time series uses the output of a function from time step t as the input for model step $t + 1$. Time series modelling involves creating a system memory by storing the new attributes and model parameters computed at each time step. Time series data can be single lists, linked to a whole map, to given points, or a stack of maps. Time series output can be time plots for single attributes for given locations, or stack of maps. Computationally, this means that the database can efficiently store and retrieve all intermediate steps in the calculations to provide feedback loops.

Stochastic variables can be generated by using Monte Carlo methods and geostatistical methods of conditional simulation (Burrough and McDonnell 1998) to generate data with a random component.

9.3   A Toolkit for Dynamic Spatial Modelling – PCRaster

Clearly, dynamic modelling with gridded data not only requires all of the usual GIS functionality but also the ability to derive surface topology and use that infor-mation to transfer data from cell to cell. All steps need to be able to be influenced by the input of times series data for every cell, efficient data storage and retrieval is needed to store and use intermediate results, and there is a need to report results to file. These files can then be displayed as time series plots and animated 2-D maps or 2.5-D drapes to provide the user with dynamic visual output.

PCRaster (Deursen and Wesseling 1995) is a dynamic modelling language that fills the gap between the standard commercial GIS and the off-line dynamic model or the once-off program. It operates in raster mode, providing a large selection of modular tools for spatial and temporal systems analysis, including a full set of mathematical tools for computing new attributes from original attributes for each cell individually. It also includes a wide range of functions for modelling spatial dispersion (neighbourhood interactions) and directed transport over topological networks. These functions can be driven by data from 1-D or 2-D time series to provide interactive dynamic models of spatial and temporal processes with feedback loops that provide outputs in the form of 1-D graphs or a 2-D stack of maps. The stack of maps can be displayed in video modes to provide realistic presentation of the results of the dynamic models. PCRaster also includes a full suite of geostatistical methods for interpolation and conditional spatial simulation (see Pebesma and Wesseling 1998).

A model written in PCRaster will adhere to the four precepts of good modelling given by Casti (1997), namely *simplicity, clarity, freedom from bias*, and *tractability*. As with conventional models a model written in PCRaster must adhere to funda-mental physical principles. Figure 9.3 provides a conceptual overview of the struc-ture of a PCRaster program, and further details are given in Box 9.1. The dynamic

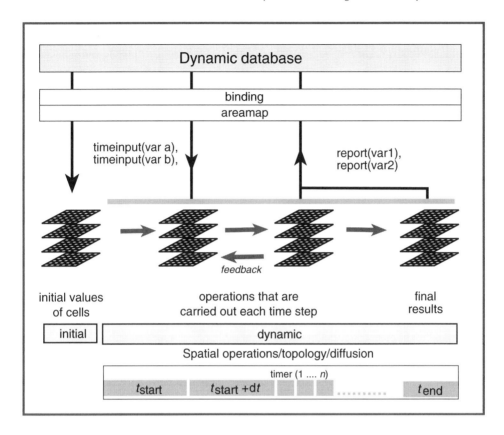

Figure 9.3 *Schematic overview of a PCRaster program*

database consists of time series and stacks of grid maps that can easily be obtained from conventional GIS, remote sensing imagery or interpolation.

Once the model has been run, the results can be displayed as static or dynamic 2-D or 3-D displays; in the latter case the display resembles a film. This enables the user to see exactly how the dynamic model has created the patterns and how these patterns change with time. By changing the model parameters and re-running the program, the effect of changing the value of single parameters, or combinations of parameters, on the results can easily be seen and evaluated, as is demonstrated in the remainder of this paper.

9.4 Examples of Dynamic Modelling with PCRaster

I give two examples, first the dynamic modelling of a non-topological dispersion process, and, second, the routing of surface runoff over directed topological nets.

Box 9.1
The structure of a PCRaster program

A PCRaster program has five main sections called *binding, areamap, timer, initial* and *dynamic*.

The *binding* section links the external file names or parameter values to the internal variable names used in the program; by simply changing file names the same model can be run with different data.

The *areamap* section defines the geographical extents of the gridded input files that will be addressed by the model – the default is the whole area.

The *timer* section specifies and controls the number of iterations for the dynamic section of the program.

The *initial* section defines the starting values of all attributes, either by directly reading from file or by creating derived attributes (e.g. the slope of a digital elevation surface) that will be used only once in the program. All of the allowed mathematical operators can be used to prepare the data in the initial section.

The *dynamic* section contains the code for all the mathematical operations for one cycle of the model. Intermediate results may be saved to time series or to stacks of gridded attribute files. The output of one iteration forms the input for the next. For example, when modelling erosion and sedimentation, each new sedimentary layer computed can be stored as a new data file so the whole set of maps forms a 3-D sediment packet.

9.4.1   The Dispersion of Plants over Continuous Space

There has been much study of the dispersion of plants and animals over the Earth's surface, particularly with respect to the role of islands and corridors in the dispersal process (e.g. see MacArthur and Wilson 1967; Johnson et al 1981; Kirkby et al 1987; McDonnell 1988). Though one can describe many different physical processes controlling the dispersal and establishment of species, in this example I limit the discussion to three independent factors, namely *dispersion, establishment* and *persistence* that govern the ability of plants to diffuse from a given starting point over a surface (e.g. a forest or landscape) and create a stable population.

- *Dispersion* is governed by the probability that seed carried from the starting point (by wind, birds, etc.) reaches location X.
- *Establishment* is governed by the probability that conditions at site X allow the seed to germinate and establish a new plant (i.e. seed is not eaten, does not rot, young plant is not killed by frost, etc.).
- *Persistence* is the probability governing the time period that the new plant survives and produces viable seed that in turn can be distributed to new sites.

Of these three probabilities, only dispersion is a truly spatial process. The probabilities associated with establishment and persistence may vary spatially and temporally (as many studies have shown) but both are essentially point processes since they do not involve lateral interactions between sites. Note that all three probabilities conceal a host of detailed mechanisms and interactions and the methods used here are only for illustration. A simple probabilistic model of dispersion ignores the details of the exact transmission of the seeds, which could be by wind, water or animals. The concepts of establishment used here ignore issues such as long-term dormancy of seed and interspecies interactions (e.g. predator–prey interactions), but they can easily be included (Deursen and Heil 1993).

In this example dispersion is modelled by an isotropic negative exponential model in which the probability of reaching a site is linked to a distance parameter governing the range to which dispersion is likely. The probability of reaching site X is given by:

$$P_i = P_p^{(\text{distance/range})} \tag{14}$$

where P_p is the prior probability of the seed reaching the distance equivalent to the range.

For establishment, P_e, the probability of a seed germinating and establishing once it reaches X is modelled very simply by comparing the probability of reaching X with the square of a value drawn from a random uniform distribution in the range 0 to 1. Using the squared value weights the probabilities downwards.

For persistence, the plant's biological activity (i.e. its active period for producing offspring at the new site) is modelled by an inverse negative exponential function of age:

$$P_d = 1 - 0.5^{(\text{AgeVeg}/2\,*\,\text{uniform(1)})} \tag{15}$$

The term 'uniform(1)' is the PCRaster operation to compute a random number from a uniform distribution. The age of the plants at each site is computed by accumulating the number of years of biological activity.

Box 9.2 lists the model code. The probabilities of dispersion to a given cell are computed using an isotropic 'spread' function embedded in a time series loop, the results of which are converted by the negative exponential model into a probability of a given cell being reached. The remainder of the loop determines whether the seed germinates and the plant establishes itself, and its biologically active life.

9.4.1.1 *Spatial Constraints for Colonisation and Establishment*

Besides functions computing probabilities of dispersal, establishment and persistence, a spatial dispersion model should also contain information about the spatial distribution and scale of those components of the landscape that are essential for supporting the species of interest. The landscape could provide suitable conditions everywhere (isotropic and homogeneous), it could contain strongly directional dispersal processes such as wind effects, it could contain islands of favourable sites surrounded by hostile terrain (isolated woods, real islands or soil conditions), or

Box 9.2
PCRaster code for simulating seed dispersion

```
# model for simulation of flower seed dispersal
# ProbReach=(O.1)^(Distance/Range)

binding
# Inputs
InitialPlants=iniscen.map;        # Map with distribution of
                                  #  plants at start of model run

Range=400;                        # Dispersal of seed
                                      # distance for which probability is 0.1

Woodland=wplain.map;              # Map with forested area given areas
                                      # that can be colonised

# outputs
Distance=distance;                # Maps with distance to nearest plants
Probability=probab;               # Maps with probability that
                                      # seed reaches cell and comes out

NewPlants=newplant;               # Maps with locations where
                                      # new plants will grow next year as a
                                      # result of seed dispersal

Plants=smplant1;                  # Maps with the distribution of plants

AgeVeg=ageplan1;                  # Maps with age of vegetation

areamap
  pladone.map;

timer
  1 15 1;
initial
# distribution of plants at start of model run
Plants = initial Plants;
# age of vegetation at start of model run
AgeVeg=scalar(1);

dynamic

# compute probability that seed reaches a given cell as function of
# distance (m) and range of dispersion function
  Distance = spread(Plants,25,1);
  Probability=0.1**(Distance/Range);

# compute probability that seeded cells yield new plants
  NewPlants = Woodland and Probability gt (uniform(1)*2);

# joint distribution of new and old plants after each cycle
  Plants=Plants or NewPlants;

# age of plants
  AgeVeg = if(Plants then AgeVeg + 1 else 0));

# death of plants – probability of dying is a function of age
  Pdeath = 1 – 0.5**(AgeVeg/2*uniform(1));
  report Plants = if(Pdeath le 0.5 then Plants else 0);
  report AgeVeg = if(Pdeath le 0.5 then AgeVeg else 0);
```

it could present continuously variable conditions with local barriers and locally favourable areas. Because all these different conditions require spatial data, the use of geographical data sources is essential.

Figure 9.4 presents the starting data and results for four simple scenarios. The initial location of the source plants is given by the map at the top of column (a). The other maps in the top row define different spatial patterns of islands where the chance of establishment is unity (everywhere else is zero). From row 2 to 5, column (a) displays some results from a simulation for which the dispersion range is 400 m and the landscape is isotropic and homogeneous; column (b) displays results for the same dispersion but colonisation is determined by seed reaching the large, widely spaced islands. Column (c) is the same as (b) except the islands are small and closer together; Column (d) is as column (b) except the range of dispersion has been increased to 500 m. All simulations were run for 15 cycles; results are reported for iterations 1, 5, 10 and 15.

Figure 9.4 shows clearly how the differences in starting conditions affect the results. In Scenario (a) the species disperses rapidly over the whole area. Once established the plant spreads everywhere and the chance of finding it at any location is given by the balance between the vacation of the cell through the death of an existing individual and the joint probabilities of new seed being deposited there and germinating.

In Scenario (b) the species disperses rapidly over its source island and spreads to the closest offshore island but disperses no further. It remains highly viable on the source area, however.

Scenario (c) shows that when the islands are close together the species diffuses rapidly over the area, reaching the target large island in the northwest. Because the islands are small, however, the species does not remain permanently established on them but depends on inward migration for the local continuation of the species.

The increased range of the species in Scenario (d) shows that it not only reaches the larger, more widely spaced islands, but travels rapidly over the area to reach the target in the northwest. Moreover, once colonised, each island is large enough to support a permanent population and the presence of the species is far less dependent on inward migration than Scenario (c).

It is a simple matter to modify the model and to build more complex dispersion models. Establishment chances need not be limited to simple binary situations but can be continuously graded according to soil, temperature, moisture and other factors to reflect habitat suitability.

9.4.2 Modelling Surface Runoff in Small Catchments

Hydrologists and soil scientists concerned with catchment monitoring and modelling need to understand the various ways that landscapes respond to temporally varying inputs of rainfall. For example, Figure 9.5 (replotted from data in De Roo 1993) demonstrates the different responses of two closely situated catchments in Limburg, in the south of the Netherlands, to a range of continuously registered rainfall events that were monitored over a period of four years. The main differences

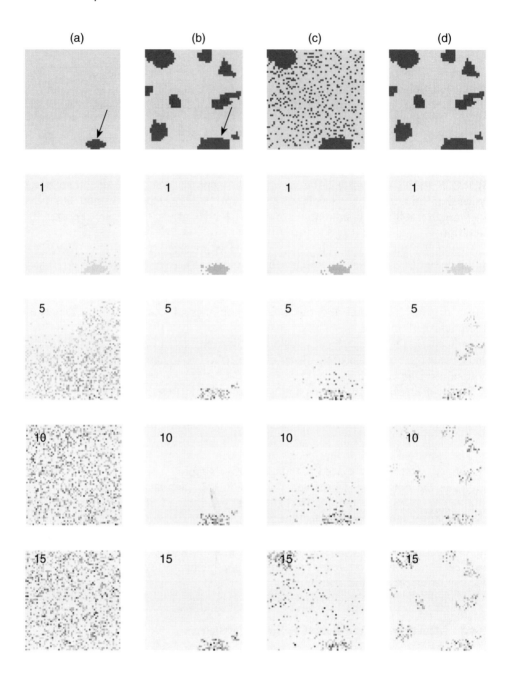

Figure 9.4 *Four simple seed dispersion scenarios*

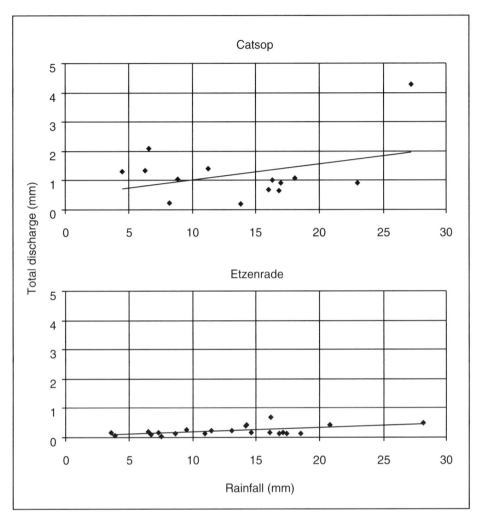

Figure 9.5 *Rainfall–runoff relations for storm events in two small catchments in Limburg, the Netherlands (1987–90)*

between the catchments are not their soil and land use, but their shape and the pattern of land use, both of which may cause some of the differences in results.

The following example illustrates the creation of a simple, distributed, dynamic surface water runoff model to investigate the problem of catchment response. The example presents data for the Catsop catchment, a 50 ha area of loess-covered rolling landscape with elevation ranging from 76 to 114 m a.s.l. (Figure 9.6). The resolution of the DEM is 10 × 10 m.

The Catsop DEM was created by thin plate spline interpolation (Burrough and McDonnell 1998) from digitised 1 m contours supplemented by surveyed spot heights. Many hydrologists (e.g. Mitasova and Hofierka 1993; Hutchinson 1995)

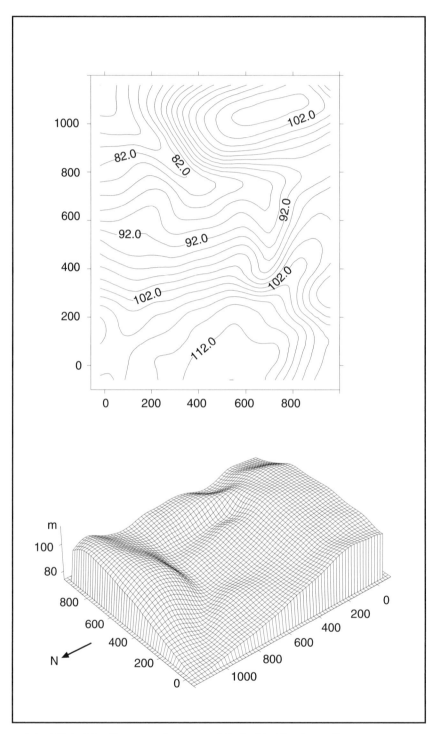

Figure 9.6 *DEM of the Catsop catchment modelled with thin plate splines*

advise the use of thin plate splines because they avoid problems of pits and topological dead ends that may occur when attempting to derive the local drain direction net from less exactly defined surfaces.

Figure 9.7a shows a subset of the LDD net (those cells having more than 40 upstream contributing cells) displayed over the Catsop DEM, and this net is used to route the excess water that accumulates in each cell as a result of inputs from rainfall and overland flow minus infiltration through the soil. The spatial pattern of infiltration is determined by the soil map (Figure 9.7b) which shows three classes of soil: sand, loam and clay, with infiltration capacities of 19, 8.1 and 2.1 mm/hour, respectively. The temporal pattern of rainfall (Figure 9.8) shows both the hourly inputs of rain for a double-front depression system over a period of 56 hours and the resulting modelled runoff time series at the outlet to the catchment.

The steps in the dynamic model are given in Box 9.3. In this example the accuthresholdflux/state operations only transmit water from cell to cell according to equations (10) and (11) when the state of the cell exceeds the local infiltration capacity. Before being reported to the database the results are converted to logarithms to aid display. Plate XI presents a selection of output maps from this model that demonstrate the variation of catchment response to the variations in rainfall and to the spatial variation of infiltration. Note that the sandy soils never cause runoff and the clay soils are always first to create discharge.

Though this is an exceedingly simple model it is capable of much extension and refinement, as has been carried out through the development of several other models like LISEM (de Roo 1996; 1998) and Rhineflow (Kwadijk 1993).

9.5   A Critical Examination of Dynamic Modelling: The Effects of Uncertainties and Roughness on Models with Special Reference to Runoff Modelling and Deposition of Sediment in Fans and Deltas

Clearly, the PCRaster approach provides an easy-to-use, quick to modify, intuitive and powerful tool for dynamic modelling, and to date there have been many scientific and practical applications. No tool, however clever, can be better than the assumptions on which it is based, and an examination of these assumptions may throw light on some current problems in dynamic spatial modelling of environmental processes.

As indicated in Section 9.2, the space–time data collected to describe natural variation are rarely completely deterministic, and neither is variation necessarily smooth and continuous. Environmental data are frequently messy, noisy and inconsistent, but often we prefer to model them by smooth surfaces rather than a proper representation of reality because the former are computationally convenient. Using smooth surfaces may yield unwanted artefacts in several situations, notably (a) the interpolation of gridded data from point observations, and (b) the derivation of a drainage topology from a smooth DEM. (See also Curran et al, this volume, for an assessment of accuracy issues of Kriged surfaces.)

Interpolation to higher resolution grids may improve visual appearance but does nothing to increase the information content of the data. When sampled data are

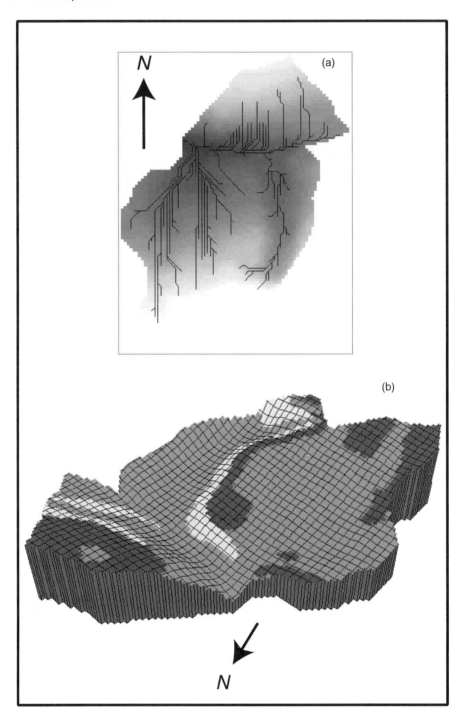

Figure 9.7   Properties of the Catsop catchment: (a) main runoff routes; and (b) soil map draped over the DEM

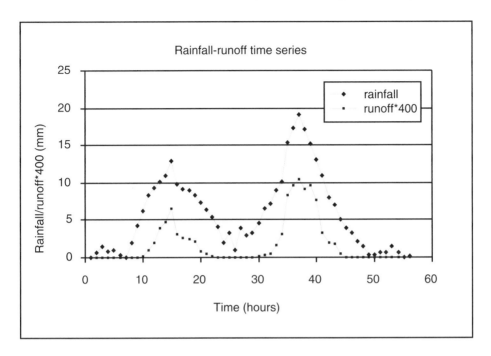

Figure 9.8   *Rainfall inputs and catchment discharge time series for the Catsop runoff scenario*

Box 9.3
Part of PCRaster code for surface runoff with infiltration

```
# add rainfall to surface water (mm/hour)
SurfaceWater = timeinputscalar(RainTimeSeries,RainZones);
# compute both the runoff and actual infiltration
Runoff, Infiltration = accuthresholdflux, accuthresholdstate(Ldd, SurfaceWater, InfiltrationCapacity);
# compute output runoff (converted to m³/s) at each timestep for selected locations
report SampleTimeSeries = timeoutput(SamplePlaces,Runoff/ConvConst);
report log of Runoff
report run = log10((Runoff + 0.001)/ConvConst);
```

---

few, increasing the grid resolution merely increases the number of cells for which the attribute values are guessed, rather than measured directly. Understanding the effects of uncertainties in the data on model outcomes is a complex exercise in Monte Carlo simulation (cf. Gomez-Hernandez and Journel 1992).

Methods used to derive surface topology from gridded DEMs may cause serious artefacts, particularly when surfaces are interpolated by ultra-smooth thin plate splines. With such mathematically exact surfaces there is only a single solution for the topology, whereas in reality small differences in surface roughness cause a diversity of flowpaths. Simple algorithms for drainage derivation from gridded smooth surfaces, such as the D8 algorithm, produce a unique solution in which the main stream line is only one cell wide. Therefore large differences may arise in the modelled upstream contributing area or wetness index between a cell on the main line and its off-line neighbour. This is counter-intuitive, because we expect cells close to each other to have similar conditions and contributing areas, especially in the bottom of valleys. Plate Xa shows an example of how deterministic algorithms concentrate the flow into a single stream line.

Linkages in topological networks created by the D8 and related algorithms are extremely sensitive to variations in the initial elevation values, particularly in areas with small differences in elevation. The sensitivity of the drainage net to initial conditions means that any single realisation of the topological net provides only one view of the drainage process. A better idea of surface water drainage can be obtained by considering the average properties of a suite of possible drainage nets that might be obtained when surface roughness is added to the DEM. The roughness can easily be modelled by a small RMS error which is added to each cell (a standard deviation equal to 0.1 per cent of the maximum relief difference in the area is enough as a first approximation); the result yields one possible realisation of the LDD. Repeating the procedure for 100–1000 times with different random values for roughness creates a probability density map of the cumulative contributing area (Plate Xb). Now the spatial pattern of water flow appears to be more realistic than when only the single LDD is used. Note that one cannot compute Plate Xb by passing a moving window smoothing function over Plate Xa.

As Plate Xb shows, computing a statistical average not only provides quite a realistic picture of the variation of upstream contributing areas over the landscape but removes the artefacts caused by the single, deterministic approach.

There are other situations where it is necessary to add random roughness to provide a realistic model of a dynamic process (e.g. Liverpool and Edwards 1995). For example, Figure 9.9a uses a deterministic diffusion equation to model the transport of sediment from the mouth of a river over an embryonic delta. There is no interaction between a packet of sediment that enters from the top centre position and the surface over which it flows, so the system has no memory.

A single diffusion equation, such as used in Figure 9.9a is of no use if we want to follow the individual trails of sediment that collectively create the delta. Figure 9.9b presents the cumulative results of 1000 simulations in which the sediment packet enters as before, but its path downhill is determined not by a general diffusion equation but stochastically by interactions determined by a LDD net for different realisations of a roughened surface. As before there is no interaction between any given flow path and previous paths so the system has no memory.

Real deltas have memories of previous events because each packet of sediment brought by the river will be deposited and contribute to the form of the delta. Initial roughness is modified not by random forces but by feedback from the sedimentation process so that for each cycle there will be a new surface for the flow. Other forces such as wave action and flooding may redistribute sediment, so removing the memory of past events.

Figures 9.9c–h present six time slices from a prototype model of delta development. In each cycle the sediment is distributed along the length of the channel

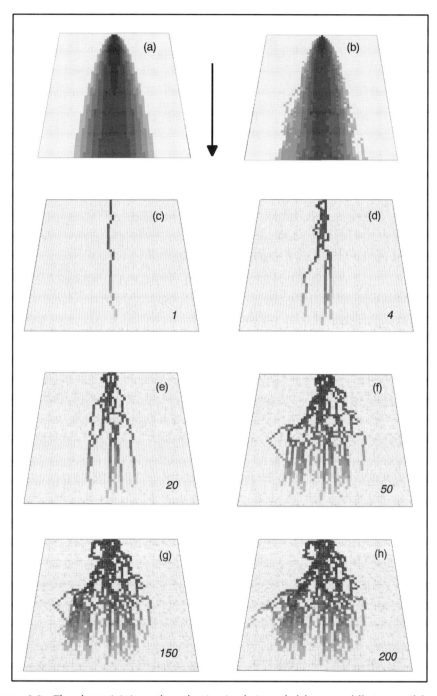

Figure 9.9   *The deterministic and stochastic simulation of deltas: (a) diffusion model – no memory; (b) 1000 realisations of randomly allocated sediment trails with no memory; and (c–h) sequential plots of stochastic models of delta formation including sedimentation (memory) and redistribution (local smoothing)*

thereby modifying the underlying surface. If the deposits are large enough the surface topology changes with each cycle.

The need for initial roughness which is modified but maintained during the development of the delta is a nice example of how a better understanding of the physical process may arise through interactive dynamic modelling because it was not anticipated before modelling began. Without the linked time series and feedback loops it is also unlikely that anyone could have predicted the kinds of characteristics that the delta deposits display during their growth.

9.6 Discussion

Besides the examples given above, many other dynamic processes can be modelled. The examples clearly show that repeating simple computations of local interactions many times leads to coherent patterns being generated naturally at a larger aggregation scale (cf. Casti 1997). The ease with which these kinds of models can be written and tested provides rich opportunities to develop ideas about spatial inter-actions and to set up testable hypotheses (Kittel et al 1996; Wesseling et al 1996).

More details can be found at *http://www.frw.ruu.nl/pcraster.html*.

References

Bates P D, Anderson M G 1993 A two-dimensional finite-element model for river flow inundation. *Proceedings of the Royal Society of London A* 440: 481–91

Burrough P A 1996 Opportunities and limitations of GIS-based modeling of solute transport at the regional scale. In Corwin A, Loague K (eds) *Application of GIS to the modeling of non-point source pollutants in the vadose zone.* SSSA Special Publication 48, Soil Science Society of America, Madison: 19–37

Burrough P A, McDonnell R A 1998 *Principles of geographical information systems.* Oxford, Oxford University Press

Casti J 1997 *Would-be worlds.* New York, John Wiley

Chatfield C 1980 *The analysis of time series*, 2nd edition. London, Chapman & Hall

Clarke K C, Olsen G 1996 Refining a cellular automaton model of wildfire propagation and extinction. In Goodchild M F, Steyaert L T, Parks B O, Johnston C, Maidment D, Crane M, Glendinning S (eds) *GIS and environmental modelling. Progress and research issues.* Fort Collins, GIS–World Books

De Roo A P J 1993 *Modelling surface runoff and soil erosion in catchments using geographical information systems.* Netherlands Geographical Studies 157. Utrecht, Faculty of Geographical Sciences, Utrecht University

De Roo A P J 1996 The LISEM project: an introduction. *Hydrological Processes* 10: 1021–6

De Roo A P J 1998 Modelling runoff and sediment transport in catchments using GIS. *Hydrological Processes* 12: 905–22

Deursen W P A, van Heil G W 1993 Analysis of heathland dynamics using a spatial distributed GIS model. *Scripta Geobotanica* 21: 17–28

Deursen W P A, Wesseling C G 1995 *PCRaster.* Utrecht, Department of Physical Geography, Utrecht University

Goméz-Hernández J J, Journel A G 1992 Joint sequential simulation of multi-Gaussian fields. In Soares A (ed.) *Proceedings of the Fourth Geostatistics Congress, Troia, Portugal.* Dordrecht, Kluwer, 5: 85–94

Hutchinson M F 1995 Interpolating mean rainfall using thin plate smoothing splines. *International Journal of Geographical Information Systems* 9: 385–404

Johnson W C, Sharpe D M, DeAngelis D L, Fields D E, Olson R J 1981 Modelling seed dispersal and forest island dynamics. In Burgess R L, Sharpe D M (eds) *Forest island dynamics in man-dominated landscapes*. New York, Springer: 215–39

Kirkby M J, Naden P S, Burt T P, Butcher D P 1987 *Computer simulation in physical geography*. Chichester, John Wiley

Kittel T G F, Ojima D S, Schimel D S, McKeown R, Bromberg J G, Painter T H, Rosenbloom N A, Parton W J, Giorgi F 1996 Model GIS integration and data set development to assess terrestrial ecosystem vulnerability to climate change. In Goodchild M F, Steyaert L T, Parks B O, Johnston C, Maidment D, Crane M, Glendinning S (eds) *GIS and environmental modelling: progress and research issues*. Fort Collins, GIS–World Books

Kwadijk J 1993 *The impact of climatic change on the discharge of the River Rhine*. Netherlands Geographical Studies 171. Utrecht, Faculty of Geographical Sciences, Utrecht University

Langbein W B, Leopold L B 1968 River channel bars and dunes – theory of kinematic waves. *US Geological Survey Professional Paper* 422L. Washington, United States Geological Survey

Liverpool T, Edwards S 1995 Modelling meandering rivers. *Physical Review Letters* 75: 3016

MacArthur R H, Wilson E O 1967 *The theory of island biogeography*. Monographs in population biology 1. Princeton, Princeton University Press

McDonnell M J 1988 Landscapes, birds and plants: dispersal patterns and vegetation change. In Downhower J F (ed.) *The biogeography of the Island Region of Western Lake Erie*. Columbus, Ohio State University Press

Mitasova H, Hofierka J 1993 Interpolation by regularized spline with tension: application to terrain modeling and surface geometry analysis. *Mathematical Geology* 25: 657–69

Park S, Wagner D F 1997 Incorporating cellular automata simulators as analytical engines in GIS. *Transactions in GIS* 2: 213–32

Pebesma E, Wesseling C G 1998 GSTAT – a program for geostatistical modelling prediction and simulation. *Computers and Geosciences* 24: 17–31

Stewart I 1989 *Does God play dice?* Oxford, Blackwell

Takeyama M, Couclelis H 1997 Map dynamics: integrating cellular automata and GIS through geo-algebra. *International Journal of Geographical Science* 11: 73–92

Wesseling C G, Karssenberg D-J, Burrough P A, van Deursen W P A 1996 Integrating dynamic environmental models in GIS: the development of a dynamic modelling language. *Transactions in GIS* 1: 40–8

10

On the Status and Opportunities for Physical Process Modelling in Geomorphology

Sue M Brooks and Malcolm G Anderson

Summary

One of the many opportunities raised from recent developments in geocomputation is the growing capability for explicit physical process modelling within a framework of digital representation of the Earth's surface. The consequence of this is an unprecedented increase in the number of 'off-the-shelf' models, available for an ever-increasing variety of model users. As stated by Longley (this volume) these models generally emphasise process over form, dynamics over statics and interaction over stimulus–response. This chapter discusses the utility of such geocomputational advances to one group of modellers, geomorphologists. It is argued in this chapter that geocomputation is important for geomorphologists, since major challenges are presented through concerns with large spatial scales in the present day (as exemplified by recent advances in GIS and digital terrain modelling capabilities), as well as extrapolation to time periods where direct observation is not possible. Process-regimes change, and uncertainty in describing an appropriate model structure and defining its associated parameter vector presents a considerable intellectual challenge. Given this, traditional geocomputational approaches require a broader base to address long-term issues, yet off-the-shelf models/geocomputational techniques may represent a dangerous distraction for geomorphological modellers, given their focus on current-day issues. Opportunities for long-term landform modelling through geocomputation encompass a set of associated issues which must be fully addressed in such research. This chapter cautions against the tendency to construct and apply models founded only on what is important in the present day, since this usually involves a move towards increasingly fine scales of resolution and increasingly detailed process representation, yet only

Geocomputation: A Primer. Edited by Paul A Longley, Sue M Brooks, Rachael McDonnell and Bill Macmillan.
© 1998 John Wiley & Sons Ltd.

limited consideration of the major model assumptions. As an example of how to overcome this in part, the chapter emphasises how *hydrological* modellers have taken up the challenge raised by the inverse problem approach, and demonstrates how this has perhaps even greater relevance for geomorphologists. In order to investigate conditions of the past which have given rise to current-day landscapes, it is necessary to interrogate model structures in some detail. This facilitates modelling of a variety of scales, and selection of the optimum temporal and spatial scale of analysis. Recent hydrological modelling advances have highlighted these issues. This chapter calls for geomorphologists engaged in physical process modelling not simply to accept and apply geocomputational tools, but to tackle the intellectual challenges raised by the inverse problem approach, to develop feedback links and process representation in models applied to the long term, and to select an appropriate scale for assessment of long-term landscape change.

10.1   Introduction

While geocomputation provides a research paradigm relevant to many disparate disciplines, the central area where it becomes relevant to the geomorphologist is in modelling long-term landform change. As such, geocomputation for the geomorphologist provides a unique set of challenges and opportunities which are not encountered in other disciplines involving model-building and application using available geocomputational techniques. Long-term geomorphological modelling is among the most demanding of disciplines for process representation and parameterisation, since there is change over the time periods of relevance. Of central concern to the geomorphologist is long-term topographic change, hence the major geocomputational input and output is the digital elevation model (DEM). Recent geocomputational advances have provided the opportunity to develop highly discretised DEMs which define detailed variation in relief through data-rich inputs and outputs. In recent years it has become possible to link such geocomputational drivers as these highly discretised DEMs to various methods of simulating change in catchment topography. These methods include distributed physically based models, neural networks, cellular automata and, more recently, lattice gas techniques (Pilotti and Menduni, 1997) for simulating sediment redistribution. However, in most examples the emphasis is firmly placed on relatively short-term topographic change. In this chapter we emphasise the use of distributed physically based modelling and its link with topographic change to explore the potential for linking geocomputational advances to the issue of long-term landform (topographic) change.

While geocomputational advances involving topographic representation have provided many novel opportunities for exploring catchment behaviour in the short term, the data demands are prohibitive for application to past time periods. We might even go as far as to suggest that this focus on detailed topographic representation represents a dangerous distraction for long-term modelling, where the focus should be somewhat different from a continued drive to adopt increasingly fine scales of data resolution. In the relative absence of complete data sources for deriving DEMs in past periods (where the picture becomes increasingly incomplete the further back we go), the tendency has been to extrapolate current datasets and process inclusion to past periods, potentially involving overextrapolation of the model. In this chapter we explore and attempt to formalise this extrapolation, in terms of data requirements and uncertainties, process representation, scale of model appropriate for long-term topographic change and opportunities for model validation. The issues are summarised in Figure 10.1.

Hence, the nature of geomorphology demands a modelling strategy which focuses on large-scale geocomputational techniques, which may include considerable change in the process drivers over the long term. Many of the more recent, relevant geocomputational advances can be found in the field of hydrological and groundwater modelling. Although modelling involving physical processes has found a place in geomorphological research in recent decades (Fernandes and Dietrich 1997; Howard 1994; Kirkby 1971; Tucker and Slingerland 1994) and several significant advances have been made concerning long-term change, there remains a divergence of approach between modelling strategies adopted by geomorphologists and those which characterise recent advances in hydrological modelling (Beven 1989; 1997; Yeh 1986). Although each is concerned with differing time and space scales, there are still important lessons which can be brought to geomorphological modelling from hydrology. This chapter aims to identify these lessons by considering recent modelling advances in both geomorphology and hydrology.

Essentially, with relatively few exceptions (some of which are listed above), geomorphological modelling has been characterised by only a limited focus on model development. Often geomorphological research has 'borrowed' models developed for small-scale hydrological applications and perhaps has developed a tendency of overdependence on such readily available models. This contrasts with hydraulics and groundwater modelling where the focus is more firmly on model development. Hence, geomorphology lacks a wide platform of readily available highly developed models appropriate for exploration of long-term system behaviour. This chapter calls for a greater focus by geomorphologists on model development issues raised so frequently among hydrological modellers.

One central concern for the application of geocomputational tools in the long term must surely be thorough consideration of the inverse problem approach, outlined by Yeh (1986). There is considerable knowledge of the outcome of system behaviour, but it is in identifying suitable model structures and parameter dimensions for previous time periods that geomorphologists face the most pressing intellectual challenge. The inverse problem is comparatively well encompassed in engineering and hydrological model developments, but remains undiscussed in the geomorphological literature. Related to this, most physics-based models exhibit a lack of identifiability, given that alternative parameter values (and model structures) may yield equivalent model performance. This issue, referred to as the problem of equifinality, has traditionally been at the heart of geomorphological research but has yet to be taken up in long-term modelling approaches. However, methods exist for optimising parameter sets (Mroczkowski et al 1997) and important advances might be gained from such approaches.

A closely related issue is the traditional emphasis on rather simplistic process classification. Where models are constructed around such process classifications, this constrains possible interpretations of system behaviour to focus only on this

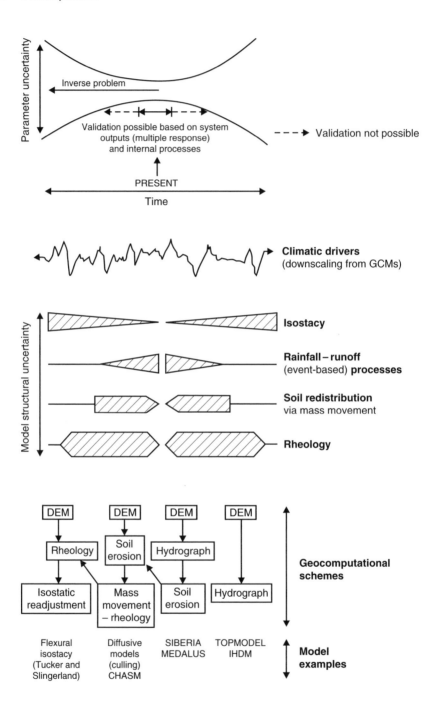

Figure 10.1 *Geocomputational issues of central concern for long-term modelling strategies adopted by geomorphologists*

subset of processes. Pilgrim et al (1978) have demonstrated that for certain hillslopes it is possible for *all* of a range of hypothesised models of hydrological response to occur, but that no *a priori* method exists to determine which response would predominate in any given situation. Examples of recent field research support this (Brammer and McDonnell 1996; Anderson et al 1997), demonstrating considerable complexity in hillslope process–response, which makes simple incorporation of neatly classified processes into geomorphological models potentially misleading (see Openshaw, this volume, for a broader discussion of multiple hypothesis testing).

As well as demonstrating the need to address geocomputational issues of process-inclusion, these examples raise concerns for scale linkages. As Wheater et al (1993) point out, small-scale behaviour can be viewed as chaotic, and thus the simple summation of process behaviour to larger scales may be inappropriate. While such variability of response characterises the small-scale, as the enquiry addresses ever-increasing scales there is a reduction in response variability. Thus, at some scale, the variance between hydrological responses for catchments of the same size is at a minimum (the representative elementary area – REA). Of relevance to geomorphologists is that recent research has shown that the REA is of the order 1 km^2. Geomorphological models which are designed to operate above this scale may be of greater utility than those which address smaller scales of enquiry.

A consequence of these factors mentioned above is that of an unhappy convergence to what Chorley (1996) has termed *engineering geomorphology*. Engineers have analytically embraced the above issues whilst, in general, geomorphologists have not. With engineering developing 'environmental' and 'softer' elements, geomorphologists may well (unwittingly) be passing the legacy of fundamental process–form research to other disciplines. Other researchers have begun to accept scenario modelling and to develop significant improvements on classical sensitivity analysis methods – so often the unfocused preserve of amateur modellers as they bypass the central issues outlined above. Thus, in full factorial experimentation all interaction effects can be evaluated (Box et al 1978). There will be a developing inconsistency in geomorphology if it augments distributed models without developing factorial numerical experimentation – effectively without attempting to decode the essential multidimensional interaction which such developments explicitly promote.

In exploring the scope for geocomputation to provide a broad base for modelling long-term landform change, this chapter distils the above arguments into four major issues which we believe need to be addressed (summarised in Figure 10.1). First, model validation related to the inverse problem needs to be considered. Validation of large-scale and long-term modelling is problematic, and the search for rigorous validation may not represent the most appropriate goal for future research in physical process modelling (De Roo 1996). Second, model interpretation aimed at establishing system behaviour must be guided by the way in which processes are simplified within the model structure. Third, the extent to which it is possible to develop scale-linkages between models needs to be emphasised. Fourth, system representation within models can be attempted at many levels of resolution, with important consequences for model outputs.

We begin by outlining modelling approaches which have characterised long-term modelling in geomorphology.

10.2   Approaches to Long-Term Modelling in Geomorphology

Of central concern when assessing long-term landscape evolution is our inability to be certain about the initial conditions which have led to the landscapes of the present day. Geomorphological modelling has progressed through a series of advances which have enabled discussion of larger-scale issues, and include increasing complexity in the representation of physical processes. Given that geomorphologists can observe in some detail these process-outcomes for present-day natural landscapes, the challenge is to minimise uncertainty in both initial conditions (parameter sets) and process combinations (model structure) which have produced these landscapes. Figure 10.1 outlines a possible scheme for tackling this challenge. With increasing levels of sophistication in physically based models, we must address the question 'to what extent can these models be used to elucidate the conditions and processes which prevailed in past periods and which have operated through time to produce our present landscapes?'

Such problem inversion is prevalent in groundwater flow and hydraulic modelling (Ferraresi et al 1996; Lunati et al 1998) – it is time to see inclusion of geocomputational techniques, such as the differential system (DS) method or Kalman Filter–based approaches, which can solve the inverse problem introduced into geomorphological models. Discussion of existing models, from the early hillslope change models (Culling 1960; 1963; 1965), right through to large-scale landscape evolution models (Tucker and Slingerland 1994; Willgoose and Riley 1997), highlights this challenge for geomorphologists and indicates future strategies for model development and application.

10.2.1   Predictive Modelling of Hillslope Change by Diffusive Processes

Numerical models for predicting hillslope change for thousand-year timescales really began with the work of Culling in the 1960s (Culling 1960; 1963; 1965). Of central concern was the adoption of a diffusion approach, whereby soil redistribution on slopes accounted for changes in elevation through time. The principles are straightforward and involve combining a slope-dependent sediment transport law (equation 1) with the conservation of mass equation for landscape evolution (equation 2), to define an equation which describes long-term elevation change in a manner similar to Fick's Law of Diffusion (equation 3). This system of equations is as follows:

$$q_{\mathrm{s}} = k \left(-\frac{\delta z}{\delta x} \right) \tag{1}$$

$$-\frac{\delta z}{\delta t} = \frac{1}{\rho_b} \frac{\delta q_s}{\delta x} \tag{2}$$

$$-\frac{\delta z}{\delta t} = \frac{1}{\rho_b} \frac{\delta}{\delta x}\left(-k\frac{\delta z}{\delta x}\right) \tag{3}$$

where, q_s = sediment flux, z = elevation, x = distance from divide, ρ_b = mass density of soil, k is a constant of proportionality, and t denotes time.

Under the assumption that the relationship between soil flux and gradient is constant along the hillslope, equation 3 can be simplified as follows:

$$\frac{\delta z}{\delta t} = D\frac{\delta^2 z}{\delta x^2} \tag{4}$$

where D is a diffusion coefficient with units (L^2/t), and is equal to k/ρ_b. Assuming diffusion is constant through time and that the boundary conditions are fixed (Figure 10.2a), it can be demonstrated that characteristic forms develop in the long term, dependent on the transport law (equation 1) in operation. These characteristic forms have been interpreted as equilibrium forms (time-independent) which differ depending only on the governing diffusive process (Kirkby 1971).

Two major limitations have been identified with this modelling strategy. First, boundary conditions are assumed constant through time (Armstrong 1987) and, second, diffusion is assumed not to vary spatially or temporally. The constant k essentially describes the ease with which soil particles are moved, and would be expected to vary as soil properties vary over space and time. Much field research in geomorphology has revealed considerable variation in soils, both spatially (Dalrymple et al 1968) and temporally (Bockheim 1980; Harden 1982; Mellor 1985).

Recently, the implications of these assumptions have been considered by Fernandes and Dietrich (1997). Their initial challenge was in parameterising the diffusivity coefficient, but they do this for present-day conditions. Fernandes and Dietrich (1997) describe a variety of methods for accomplishing this, and emphasise a numerical solution to the slope–distance relationship derived for steady-state conditions, where elevation change ($\delta z/\delta t$) is constant and equal to basal down-cutting. Substituting this relationship into equation 4 and integrating with respect to distance produces the following relationship between gradient and distance:

$$\frac{\delta z}{\delta t} = \frac{b_d}{D}x + C_1 \tag{5}$$

where b_d = downcutting rate at base of slope and D = diffusion coefficient. With flat divides, $C_1 = 0$. The gradient of the slope–distance relationship gives (b_d/D), so by measuring basal downcutting it is possible to find a value for diffusivity. The

(a)

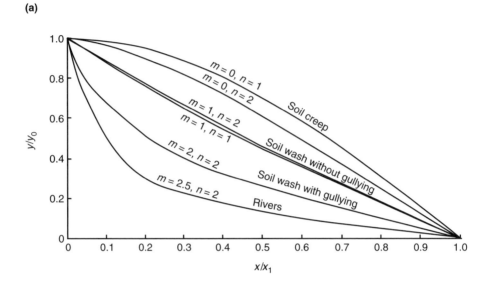

(b)

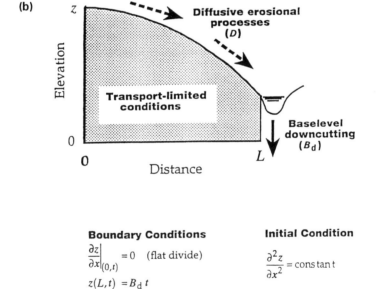

Figure 10.2   *Hillslope process–response modelling. (a) Characteristic slope forms for fixed divide and base level conditions (after Kirkby 1971). (b) Scheme for modelling hillslope development with varying boundary conditions (Fernandes and Dietrich 1997)*

estimates produced by this and other methods range from 4 to 400×10^{-4} m^2/year. It is then necessary to parameterise the downcutting term, and this is achieved using field-measured data for landscape lowering, assuming that this is equivalent to basal downcutting (Young and Saunders 1986). Values for downcutting range from -10^{-5} to -10^{-2} m/year. Convex hills are estimated to range from 25 to 100 m in length, and the model is then run for a range of parameter combinations (see Figure 10.2b). Under increased basal downcutting (twofold), hillslopes become steeper, hillslope curvature and erosion rate increase through time, and total sediment flux increases to a new equilibrium value. The results suggest that the time for attainment of equilibrium is of the order 90 ka. The model was then used to examine how slopes behave under changes in basal downcutting and diffusivity. Initially, a doubling of diffusion produces slope profiles which flatten (lower curvature) and involve progressively declining sediment flux until a new equilibrium is reached after 70 ka. The final part of this modelling exercise involves applying step changes in diffusivity and basal downcutting to establish the time taken to reach a new equilibrium (relaxation time). The results are shown in Figure 10.3 and indicate (a) the longer the hillslope the greater the relaxation time; (b) when diffusivity or basal downcutting is increasing this results in decreasing relaxation times with greater degrees of parameter change; and (c) decreasing diffusivity or basal downcutting results in lengthening of relaxation times. The model is far more sensitive to diffusivity than to basal downcutting.

This modelling approach, wherein a series of equations is used to simulate long-term hillslope change, has been central to geomorphological modelling. It does possess certain distinct advantages, especially in its simplicity to implement and to parameterise. The parameters are spatially and temporally lumped, and there are just three (slope angle, basal downcutting and diffusivity) to consider. The model also contains a number of empirical constants which can be varied for different scenarios, but which only affect the rate of change rather than alter fundamentally the detailed process mechanics. Hence the approach has been used with some success to address a central question in geomorphology concerning whether or not landscapes attain steady-state and over what period. Field evidence alone provides contradictory views, with some research supporting time-independent forms (Hack 1975), while other evidence supports the idea that the landscape is a palimpsest of forms reflecting differing past climates (Chorley et al 1984). Relaxation times influence landform persistence and from this modelling advance we see relaxation times spanning 10 000 to 10 000 000 years. That these exceed the frequency of climatic oscillations for the last few million years (Shakleton et al 1988) suggests that the landscape may well contain a legacy from several past climatic phases. Modelling is beginning to provide answers to such central concerns for the geomorphologist, without necessitating a detailed focus on process mechanisms. However, this rather avoids certain other issues (see Figure 10.1) which need to form a focus for discussion among geomorphologists.

Models based on these diffusivity principles provide no insight to process mechanisms. Change is a function of slope gradient, the rate being determined by empirical constants (see Figure 10.2a). The two other parameters, basal downcutting and diffusivity are treated as being spatially and temporally lumped.

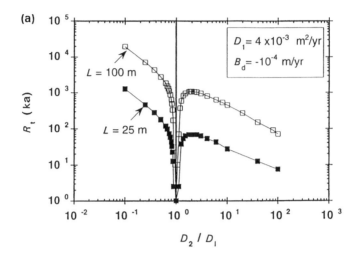

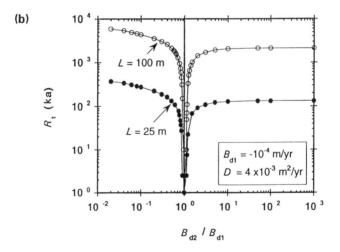

Figure 10.3 *Relaxation times associated with step changes in (a) the diffusion coefficient; (b) the base level downcutting rate. Slopes are 25 m or 100 m in length. The right hand side shows increases in parameter values; the left hand side shows decreases*

Fernandes and Dietrich (1997) attempted to examine the implications of variation in these two parameters, but in doing so they demonstrated that the two parameters were themselves linked. Furthermore, they carried out a sensitivity analysis on the parameters, but did not allow for possible temporal change in parameter values or, in the case of diffusivity, variation over the hillslope. They viewed the problem as one needing derivation of accurate diffusivity and basal downcutting for a specific field site, and then ran the model with these values to establish relaxation times for different parameter values. What is really required to understand system behaviour is to apply more complex models which take account of a variety of processes, have a greater number of distributed parameters, include methods to assess associated uncertainty and allow for change in parameter values and process operation through time. With known outcomes, the problem is to define the optimal parameter set with minimum uncertainty as to which gives rise to this particular outcome.

10.2.2 Reconstructing Parameter Sets for Process Modelling for Past Periods

Hillslopes evolve through rapid mass movement when they are steep or composed of relatively weak material (Carson and Petley 1970; Young and Saunders 1986). When a slope fails it has a factor of safety less than or equal to unity, where the downslope shear force exceeds the material shear resistance. This fact has been recognised by engineers for decades, who design slopes to remain stable for particular slope and material properties. A 'classic' approach to explaining why slopes fail is to carry out back-analysis (Skempton 1964). In this, the factor of safety is known (i.e. it is equal to unity) but the parameters which produced this outcome are not. Different combinations of parameters may combine to cause slope failure, and the challenge lies in ascertaining which particular set was responsible.

In the Holocene (the period spanning approximately the last 10 000 years), phases of slope instability have been identified by dating buried soils (Brazier and Ballantyne 1989) or by measuring past sedimentation rates in lakes (Page and Trustrum 1997). Various explanations have been offered for these phases of instability. In North West Europe climatic oscillations may have provided higher rainfall intensities, totals or durations, and this is frequently suggested as a cause of slope failure in the past (Grove 1972). Similarly, in New Zealand, increased cyclone activity related to El Niño variation has been suggested as a cause of periodic slope failure. However, for certain time periods vegetation change might provide a different explanatory mechanism (Innes 1983; Page and Trustrum 1997). It is well known that soil profiles change in their geotechnical and hydrological behaviour over time (Brooks et al 1993). In accounting for slope failure in the past, all three explanations are plausible.

Recent model developments and their application in geomorphology have enabled this problem to be addressed to a certain extent. Models which include small-scale soil physics principles (Hillel 1980) have enabled the soil characteristics which might give rise to slope failure to be ascertained. Developing soil profiles undergo changes in depth, morphology and internal differentiation, some of which are significant to slope failure analysis. The precise changes can be quantified by examining land surfaces of different age, but with similar parent material, topography and climate (Jenny 1941) and there are well-documented examples for soil profiles of different character (e.g. Harden 1982; Mellor 1985; Robertson-Rintoul 1986). Using similar modern-day analogues, climate change effects can be considered (Brooks and Richards 1994). Rainstorm characteristics associated with different present-day synoptic conditions (e.g. cyclonic or anticyclonic) can be established and, combined with information on changes in synoptic conditions in the Holocene (Lamb 1982), temporal changes in rainstorm characteristics can be estimated. Finally, vegetation variation can be reconstructed through pollen analysis (Page and Trustrum 1997), making it possible to define the likely range of parameter values which prevailed at different times in the Holocene. To investigate the combined effect of rainstorm type, soil development and vegetation change on hydrology and long-term slope stability, a 'snap-shot' approach is required.

The model which is used here to illustrate this approach is a physically based finite-difference Combined Hydrology And Stability Model (CHASM), although other models are available with which it might be possible to adopt a similar approach. The detailed process-basis of CHASM, along with its associated data requirements, are well described in various papers (Anderson and Howes 1986; Anderson et al 1988; Brooks et al 1993), and the 2-D version is depicted in Figure 10.4. In terms of current-day applications, CHASM has relatively data-rich demands and detailed process representation. The model simulates hydrological response to individual rainstorms or several storms in a series, in which precipitation is input on any selected time base. The geometry includes soil profiles defined at very high spatial resolutions (a matter of centimetres), both vertically and laterally. This is essential if the hydrological dynamics of developing soil profiles are to be investigated (see below). Finally, vegetation is included as a canopy store, combined with elevated hydraulic conductivities and soil shear resistance in the rooting zone (Brooks et al 1995; Brooks and Collison 1996). Moisture redistribution is simulated over the iteration period (usually seconds), according to the Richards Equation for saturated and unsaturated flow. At the end of each hour the model performs a stability analysis, describing the shear surface as either infinite planar, circular or non-circular. For this, soil shear resistance is described by the Mohr–Coulomb Failure Law, with direct inclusion of pore water pressures.

CHASM can act as a 'stand-alone' hillslope stability model, where detailed representation of parameter uncertainty makes it particularly useful for assessing slope stability in previous time periods, and the associated likelihood of topographic change. Alternatively, it can be utilised in wide-area analysis by coupling with SPANS GIS, where the data demands are somewhat more elaborate. The use of this second approach is described in Anderson et al (1996), where the topographic data demands are met by the use of digital video geographic (DVG) surveys (airborne laser terrain profilers with vegetation penetration). Such methods enable the creation of a DEM for difficult and inaccessible terrain, and coupling with CHASM enables the stability of each land surface unit to be assessed. Hence the major output involves a series of detailed maps of stability conditions. The full scheme is shown in Figure 10.5.

NB: Slope angle and form is governed by cell geometry

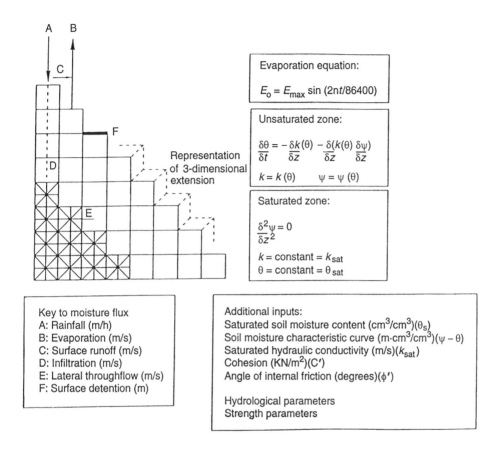

A B

C

F

Representation
of 3-dimensional
extension

D

E

Evaporation equation:

$E_o = E_{max} \sin (2\pi t/86400)$

Unsaturated zone:

$$\frac{\delta\theta}{\delta t} = -\frac{\delta k (\theta)}{\delta z} - \frac{\delta(k(\theta)}{\delta z}\frac{\delta\psi)}{\delta z}$$

$k = k (\theta) \qquad \psi = \psi (\theta)$

Saturated zone:

$$\frac{\delta^2\psi}{\delta z^2} = 0$$

$k = \text{constant} = k_{sat}$
$\theta = \text{constant} = \theta_{sat}$

Key to moisture flux
A: Rainfall (m/h)
B: Evaporation (m/s)
C: Surface runoff (m/s)
D: Infiltration (m/s)
E: Lateral throughflow (m/s)
F: Surface detention (m)

Additional inputs:
Saturated soil moisture content (cm^3/cm^3)(θ_s)
Soil moisture characteristic curve ($m\cdot cm^3/cm^3$)($\psi - \theta$)
Saturated hydraulic conductivity (m/s)(k_{sat})
Cohesion (KN/m^2)(C')
Angle of internal friction (degrees)(ϕ')

Hydrological parameters
Strength parameters

Figure 10.4 *Two-dimensional version of the combined soil hydrology and slope stability model (CHASM) (Brooks et al 1993)*

While such geocomputational advances are available to the geomorphologist, the acquisition of such detailed spatially referenced topographic data is impossible for past time periods. With respect to 'snap-shot' modelling for past time periods, of greater relevance is the inclusion of stochastic variation in parameter sets to account for the effect of parameter uncertainty through time. To accommodate stochastic variation CHASM requires that a distribution is specified for each parameter by its mean and variance. Multiple simulations are then carried out, with each individual simulation involving a different value for each parameter drawn randomly from the specified distribution. The resulting output involves a factor of safety distribution, where the probability of obtaining a factor of safety of unity (i.e. indicating the occurrence of slope failure) can be ascertained. Hence the effect of parameter uncertainty can be partially included. Figure 10.6 shows some typical results, involving long-term change in slope stability related to each of these possible

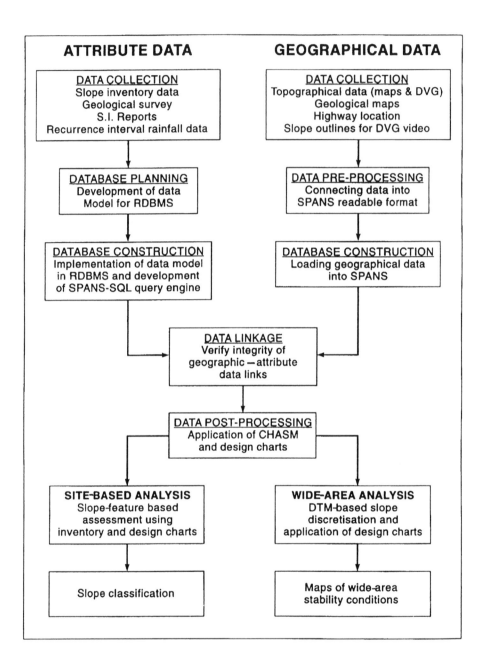

Figure 10.5 *Overall structure of the site-based wide-area CHASM through SPANS GIS (Anderson et al 1996)*

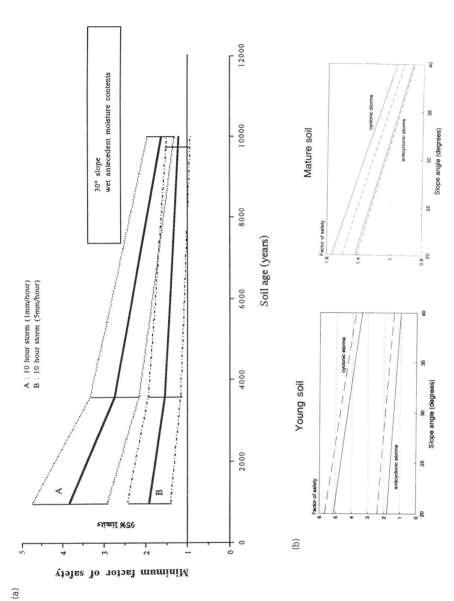

Figure 10.6 *Output from the soil hydrology–slope stability model showing the varying effect of soil profile development, storm type and vegetation on the factor of safety (solid lines in lower graphs indicate grass-covered slopes and broken lines indicate forested slopes (Brooks et al 1993)*

controls. The soil profile changes show a strong influence on slope stability, whereby in the example shown, several thousand years of pedogenesis is required for slope instability, all other factors being equal (Figure 10.6a). Soil profiles appear to exhibit varying sensitivity to climate (Figure 10.6b), although it is the higher intensities characteristic of anticyclonic conditions which are the more likely to promote slope instability. Finally, the influence of vegetation is shown (Figure 10.6b), with a change from forested to grass slopes giving rise to a greater likelihood of failure.

Models such as CHASM thus provide a basis for the evaluation of the effect of a specified set of conditions upon the hydrological and geotechnical dynamics of a hillslope. While such schemes have certain utility in the discrete analysis of parameter sets, we must caution against the extensive temporal extrapolation of such schemes in terms of long-term climatic contexts. In such contexts, new and critical factors become relevant, such as models of past failure mechanisms, which include adaptive meshing and the specific incorporation of process links that serve to redefine both the boundary conditions over time and the within-mesh parameter set. These factors, together with apparently opposing needs to have models that are mass conservative (small time step) and yet capable of operating over relevant climatic periods (long simulation runs), explain in part the pressures to make 'event'-based analytical solutions 'work' for inappropriate circumstances.

10.2.3   Focusing on Modelling Topographic Change over Time

The above two examples relating to the diffusive approach and event-based models illustrate sharply the need to construct models that, quite literally, capture the middle ground. Such new model formulations may well take the output from an event-based model (especially to obtain indications of process dominance) as effective input into model design for the longer term (e.g. the need to model climate–vegetation links with slope instability or the inclusion of rheological behaviour of soil). Such new models should capture anew elements that are critical in the long term, but are currently excluded for reasons of convenience. For example, post-failure mechanisms and resultant topographic redefinition are currently not included in models of hillslope change. Finally, the new models should have a focus on form and be prepared to be integrative and consumptive of process details to achieve such a goal – a stance taken *par excellence* in the case of the Culling style of models.

Over time, the above models indicate that considerable topographic change is possible, whether due to changed rates of downcutting or soil diffusivity, or to the occurrence of shallow translational or deep-seated rotational slope failure. Including topographic change in geomorphological models requires inclusion of feedbacks on different timescales. Over decades, redistribution of soil within catchments alters water flow pathways and positioning of convergence zones, which then alters position of erosion source areas (Beven 1997; Willgoose and Riley 1998). Over thousand-year scales, continued erosion leads to isostatic readjustment which, in turn, affects relief and subsequent erosion potential (Tucker and Slingerland 1994; 1997). However, to address these feedbacks adequately, the behaviour of sediment in motion should be included as this determines areas of deposition. It is not sufficient simply to consider changes in topographic gradient in accounting for sediment deposition; the rheological behaviour of the sediment–water combinations needs to be modelled (Iverson 1986). A few recent model developments which address these issues are shown in Figure 10.1, with greater detail being provided in Figure 10.7.

Redistribution of sediment over thousand-year timescales is tackled by Willgoose and Riley (1998) using the SIBERIA model (Figure 10.7a). This modelling scheme offers considerable advantages over earlier attempts to model soil erosion, such as the Universal Soil Loss Equation (USLE) or CREAMS (Chemicals, Runoff and Erosion from Agricultural Management Systems) models, which calculate long-run sediment generation which is lumped in time. The SIBERIA modelling framework permits topographic change in the catchment which then feeds back to determine flow convergence zones. These then become source areas for subsequent erosion. By including these feedbacks, the scheme can model changes in erosion source areas over time. The utility of such modelling schemes in the short term (10–100 year timescales) can be considerable, but application of such models in geomorphology raises concerns about process representation. Topographic change results from sediment redistribution, but over longer timescales (1000–100 000+ years) the feedback mechanism in which sediment redistribution affects topographic change is excluded. Hence, rather than trying to assess long-term change through detailed evaluation of the processes and locations of sediment erosion, a more lumped approach might well provide information about sediment redistribution at a scale appropriate for modelling such feedbacks. In such schemes sediment sinks are not included, although the consequent sediment loading on different parts of the region provides an important mechanism for topographic adjustment to occur.

In the long term a completely different emphasis on sediment redistribution processes is required. Treatment of this feedback, involving flexural isostacy (or topographic readjustment to differential loading) has been handled in 1-dimension by Tucker and Slingerland (1994), as shown in Figure 10.7b. Here sediment removal from the escarpment crest is compensated for by isostatic uplift, which extends landward towards sites of greatest denudation. This increases total relief between the escarpment crest and base level, which increases stream power and erosion rates. Furthermore, the increase in gradient on the plateau makes it harder for larger streams to capture smaller streams, and so patterns of erosion are 'equalised'. Field evidence of existing escarpments supports this result.

Finally, given that sediment redistribution and topographic change is so critical and that feedbacks are being introduced into geomorphological models on several scales with respect to several different effects, likely areas for sediment redeposition need to be treated carefully. Many models adopt the idea that deposition occurs where slope gradient is reduced below a critical threshold. However, the behaviour of sediment–water combinations when in motion is complex, and depositional areas are not straightforward to locate. The modelling of Iverson (1986) represents an important contribution in this respect, since it involves the behaviour of sediment in motion acted on by varying stress fields. Such modelling, while developed for individual hillslopes evolving by mass transfer of sediment, offers potential for coupling with isostatic flexuring ideas as sediment is moved around.

(a)

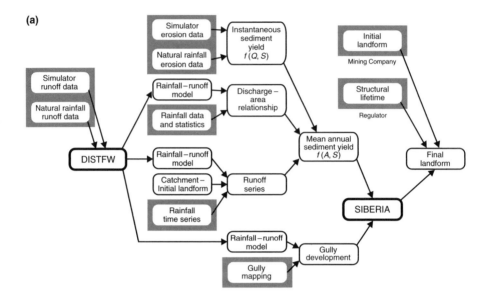

(b)

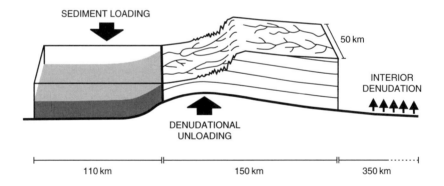

Figure 10.7 *Models available to simulate long-term feedback effects in: (a) soil erosion, hydrology and topography (Willgoose and Riley 1998); (b) isostatic adjustment and denudation with flexural upwarping (Tucker and Slingerland 1994)*

Geocomputational issues emerging from the preceding discussion of relevance to long-term modelling largely relate to the fact that opportunities for parameterisation and validation are only available for a very short time frame (see Figure 10.1). Perhaps of greatest importance is the need for major advances in geocomputational techniques to solve the inverse problem. Currently, parameterisation for past periods is achieved through field measurement using modern-day analogues. However, there are opportunities, given the known system outcome, for considering which parameter combinations give rise to a given outcome and the associated uncertainty in parameter vectors. This is particularly problematic, since over timescales of significance in geomorphology there are normally wide-ranging changes in system morphology and boundary conditions. The examples involving changing topography can be used as illustrations of this. Model structure and calibrated parameters which give a good fit between model and measured outcome for one topographic representation of a catchment may not apply in subsequent time periods where morphology has changed. In the SIBERIA model, calibration is based on present-day field experiments under simulated and natural rainstorms. The model is then used for thousand-year simulations. As soils change, one would expect that these calibrated values will become increasingly unreliable as estimators of catchment erosion. The related issue of process inclusion also requires further discussion. Successful and reliable parameterisation depends on models having adequate process inclusion, against which system behaviour can be evaluated. Finally, the issue of scale deserves close attention. The inclusion of small-scale physics has aided understanding of mass movement process-controls for example, but when compared with the modelling scheme of Tucker and Slingerland (1994), all we really need to know is 'how much sediment?' and 'where is it going to?' to parameterise a model which includes the feedback between topography and tectonics. In such cases, we need to question the extent to which such small-scale physics sheds light on large-scale responses. These are the important issues for geomorphologists concerned with modelling long-term change and represent significant challenges which must be tackled using the latest geocomputational advances.

10.3 Future Modelling Issues of Relevance in Geomorphology

The main issues identified above are those of model validation and solutions to the inverse problem, process inclusion, and the scale appropriate for application of models in geomorphology. Recent research in hydrological and groundwater modelling has highlighted each of these issues, as this section will demonstrate, providing a possible link between geocomputational techniques and geomorphological research, and suggesting a framework within which we can interpret the results of model applications in the long term.

10.3.1 Model Validation and Solution to the Inverse Problem

There has been a considerable change in attitude among hydrological modellers in recent years concerning appropriate validation strategies, why validation is carried out and whether it is necessary to validate models at all, given difficulties raised by long-term and large-scale applications (De Roo 1996; Mroczkowski et al 1997). Calibrated parameter sets which provide acceptable model outputs when compared with measurements are frequently assumed to be those which continue to govern system behaviour over future time periods. In hydrological modelling, the timescales are comparatively short, but in geomorphology we are increasingly involved with the long term. One of the central issues in hydrological modelling is how a system will behave under changing land use, but given the longer time periods for geomorphological modelling, even greater change can be expected in the system, and issues related to defining a calibrated parameter set with acceptable validation assume even greater significance. Model validation is traditionally achieved by comparing field measurement of outputs against model simulated results, but recent research has indicated the considerable uncertainty associated with this. Different validation strategies which may have relevance to the long term will be discussed below.

10.3.1.1   *Split-Sample Validation Based on a Single Output*

Powerful validation is able to discriminate between good and bad model formulations, but this depends on the type and quality of data available. Discharge records are frequently the only source of validation data for catchment hydrological models, hence their widespread use. In traditional validation exercises it is common to use a split sample test, in which a period of observation is used to calibrate the model parameters, followed by a period in which the calibrated model is used to make predictions. These are then compared with observations to examine how well the data fit. For hydrological models it is usually discharge at the catchment outlet which is employed (Abbott et al 1986). However, recent research has demonstrated that the split-sample validation exercise based on discharge records only works well in a limited number of situations, and it has even been suggested that close correspondence between model predicted discharge and that measured in the field is a *minimum requirement* for testing model performance (Kuczera et al 1993). If model performance is tested using streamflow data it can be shown that lumped, semi-distributed and distributed models perform equally well (Refsgaard and Knudsen 1996). Furthermore, Mroczkowski et al (1997) point out that in split-sample techniques, model performance is validated against the same output to which it was calibrated which gives limited opportunity for model filtering. Hence, if models are to be used to explore system behaviour and to discriminate between different model formulations and parameter sets, the split-sample test is inadequate.

10.3.1.2   *Multiple Response Validation*

Many geomorphological models are concerned with several outputs. The SIBERIA model requires successful simulation of runoff, erosion and gully formation, and the MEDALUS model simulates landscape evolution for changing hydrology and erosion, under variations in land use. In trying to use a modelling strategy to explain the processes accounting for changing catchment response in disturbed catchments, Mroczkowski et al (1997) adopt two different model formulations. They have available three datasets spanning a ten-year period for (a) runoff, (b) chloride concentration in the stream, and (c) average groundwater levels in an undisturbed and a clear-felled catchment. A series of calibration and validation exercises is carried out for each model formulation applied to each catchment. The first simulation applies to the undisturbed catchment (Figure 10.8a), involving multiple response in all three variables. The model is calibrated against all three responses to find the global parameter optimum. This parameter vector is then used to simulate catchment response. Examination of residuals shows that for all three responses, predictions lie within 90 per cent error bands, implying that this particular model formulation is an adequate representation of catchment response.

The same model formulation was then applied to a disturbed catchment, with similar hydrological characteristics. Initially, a split sample single response calibration and validation exercise was carried out, with the calibration period being for the first 36 months of record. For the single response (runoff) the residuals all lie within the 90 per cent prediction limits, suggesting that this model formulation is appropriate. However, when a multiple response strategy is applied to the disturbed catchment, predictions for chloride concentrations and, to a lesser extent, groundwater levels, plot outside the 90 per cent limits (Figure 10.8b). The parameter vector identified in the calibration phase cannot consistently simulate all three responses with the implication that this model formulation is not an adequate representation for disturbed catchments.

For geomorphological modelling where parameter values have changed over time, calibration for one unique set of parameters is impossible. Even where parameter vectors have not changed, successful calibration for one response (e.g. runoff generation) may not necessarily produce an equally successful simulation of other responses (e.g. sediment production). Various validation strategies have been suggested and these are discussed in the following section.

10.3.1.3   Internal Validation

Physical process modelling has frequently focused on matching the predicted final outcome with measurement. With the CHASM model, this final outcome is the factor of safety. When it falls below unity failure will occur. Hence the only validation opportunity occurs for slopes which actually fail. For slopes which have yet to undergo failure, the factor of safety is unmeasurable. The CHASM model does, however, offer opportunities for validation against variables operating at a more detailed level of resolution but which control the likelihood of failure. Such validation has been termed internal validation (Fawcett et al 1995), and places lesser emphasis on final model outcome. This approach has been supported by Mroczkowski et al (1997) by adopting a second model formulation, with more detailed process representation in the near-channel zone. Multiple response validation for the disturbed catchment with this model was much better than with version 1 (where fewer processes were included), and the authors accepted that this

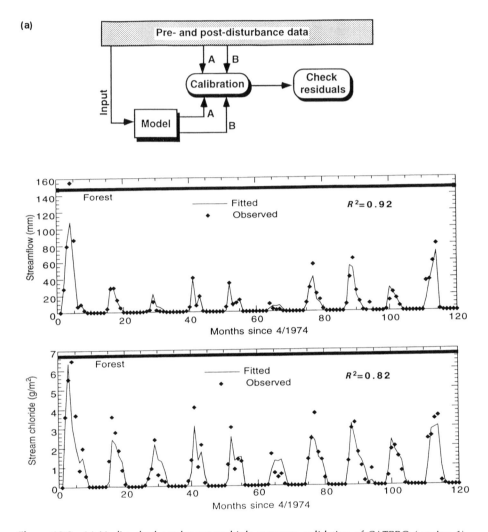

Figure 10.8 (a) Undisturbed catchment multiple response validation of CATPRO (version 1)

more detailed model formulation was probably an acceptable simulation of system behaviour. They continued with a further independent validation exercise involving comparison of the internal response in observed and predicted saturated areas. Again, predictions lay within the 90 per cent limits and the model formulation was felt to provide a good representation of the hydrological response in the disturbed catchment.

Research in hydrological modelling outlined above suggests two important new strategies for model validation, one involving multiple response calibration and validation, while the other involves internal validation. It also demonstrates that split sample exercises based on stream discharge do not necessarily test the reliability of physically based models for application under land use or climate change

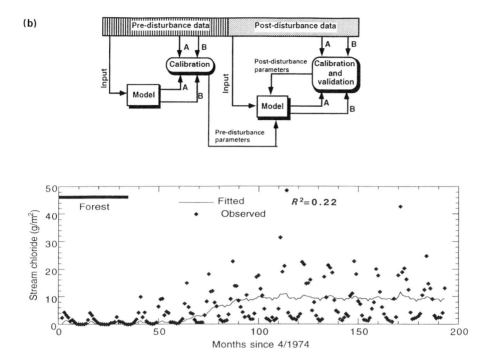

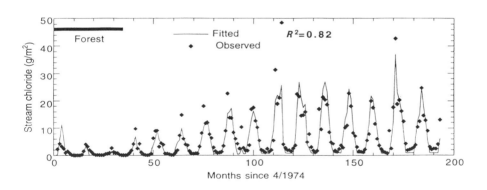

Figure 10.8 *(b) Disturbed catchment multiple response validation and results for stream chloride using versions 1 and 2 of the CATPRO model (Mroczkowski et al 1997)*

scenarios. This is of concern given the central role of land use change in hydrology and system change in geomorphology (Beven and Kirkby 1979). Other recent papers discussing model validation also suggest a strong need for internal validation, such as for TOPMODEL, where validation is often based on the dynamics of the saturated area rather than outlet discharge. This procedure might well be made easier given recent advances in remote sensing as discussed by Bates et al (1997), where floodplain inundation is simulated and compared with remotely sensed observations.

More controversially, recent research has questioned whether validation exercises are actually worthwhile, or even possible at all, given scale differences between measurements and predictions (De Roo 1996), as well as limitations in the structure of many models (Anderson and Brooks 1996). This clearly has implications for geomorphological modelling where a complex range of landforms results. Deriving simple landform parameters against which any given model prediction can be tested is far from straightforward, but nonetheless represents an important future line for physical process modelling in geomorphology.

10.3.2 Process Representation

In respect of long-term modelling two questions can be raised which require attention, related to process representation. The first of these concerns the fact that many process equations have been derived for small-scale situations and are then applied to large-scale situations. A good example is Darcy's Law which defines a linear relationship between flux and hydraulic head, which forms the basis for many hydrological models used under the 'snap-shot' approach described previously. This law was derived under controlled laboratory conditions, using small samples and taking a relatively low range of hydraulic heads. It therefore is not a good descriptor of large-scale flux under variable hydraulic heads, especially when the latter reach high values and flow becomes turbulent. The second question involves changing process dominance in the long term, where for different timescales a different combination of processes might account for slope change. In geomorphology there has been field recognition of the increasing complexity developed in landscapes over time (Dalrymple et al 1968). Recent field studies highlight this complexity and sound cautious notes for process inclusion in 'physically based' geomorphological models. Over even greater timescales, the role of the rheological behaviour of material, as well as the interaction between topography and tectonics become critical and simple hydrological models are inadequate.

The extent to which it is possible to seek physical meaning from the outputs of 'physically-based' models has been frequently questioned (Beven 1989; 1993; 1997), particularly where it is necessary to utilise calibrated parameter sets which provide unrealistic combinations of parameter values. Beven's (1997) arguments are based largely on experience with TOPMODEL, a 'quasi physically based' hydrology model which uses topography to drive the hydrological response, but similar caution is appropriate to the use of geomorphological models. Certain models used by geomorphologists, such as IHDM (Institute of Hydrology Distributed Model), include Darcian flow to describe water movement through porous media. Pipeflow (macropore flow) is omitted, and in catchments where pipeflow dominates it is only possible to gain realistic catchment responses by using 'effective' values for saturated hydraulic conductivity, which are normally larger than those measured for the matrix. In geomorphological models grid elements are frequently large and representative values for hydraulic conductivity need to combine with Darcian flow principles to provide realistic runoff generation. This simple example illustrates how a simplifying process assumption can be combined with suitable parameterisation to provide good simulation. Neither is realistic, yet the net effect is acceptable. Poor process representation can be accommodated in the parameterisation, hence the need for full consideration of model structural assumptions.

With increasing landscape complexity, geomorphological models should consider a wider range of processes than those which appear to be the currently dominant processes. In the CHASM application to the long term, soil development formed an initial focus (Brooks et al 1993). Treatment of vegetation was absent from this study but soil change was shown to be important to slope failure. However, where vegetation is present, the depth and type of slope failure can change and, in such circumstances, progressive soil development may be comparatively less significant. More recently vegetation treatment has received close attention, not just in CHASM but in a range of other models, such as the SWATRE (Soil Water Actual Transpiration Rate – Extended) and SVAT (Soil–Vegetation–Atmospheric Transfer) models (Brooks et al 1998; Wesseling et al 1998; Wigmosta et al 1994). The latter sets of models provide a more detailed treatment of the role of vegetation, in particular including subroutines for removal of water from the soil by transpiration. In certain circumstances, notably for shallow translational slope failures, or for failures in deep, weathered parent material, maintenance of soil suctions frequently promotes slope stability (Brooks and Anderson 1995), with vegetation potentially being instrumental in maintaining high suctions. Hitherto, process representation for vegetation in slope stability models has been considerably simplified (Brooks et al 1995), involving no water uptake, uniform rooting zone character (for permeability and reinforcement effects), and inappropriate description of the failure surface.

Recent field studies highlight the issue of complex hillslope processes further, whereby hydrological models have been shown to provide inadequate representation of hillslope hydrological response. Again, with landscapes becoming increasingly complex over time as soil is redistributed, the processes which govern catchment response to rainfall need to be fully included in models. Results presented by Brammer and McDonnell (1996) have emphasised the role of bedrock microtopography in influencing chemical and physical components of stream response to rainfall (Figure 10.9a). As the water-table falls below the bedrock–soil interface, pockets of water become isolated in microtopographic depressions. These water stores involve long residence times, affecting chemical composition, but also provide sources of antecedent moisture which can readily be mobilised during subsequent storms. This field exercise clearly demonstrates that we must not continue to treat the soil–bedrock interface as a smooth planar surface, but that models must take account of microtopographic variation to provide proper representation of system behaviour. A more recent example takes this further by providing a scheme for water movement which is not adequately accounted for in physically based hillslope hydrology models (Anderson et al 1997). Figure 10.9b shows a two-part flow system comprising rapid subsurface water movement through fractured and weathered bedrock, combined with slower vertical

(a)

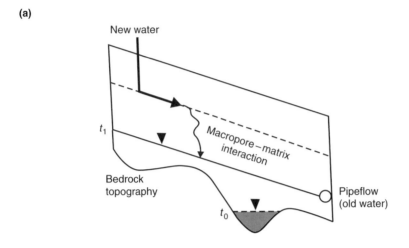

(b)

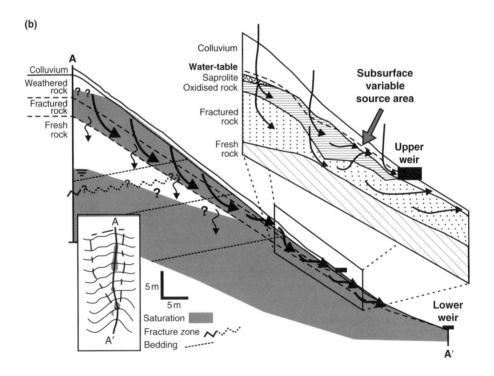

Figure 10.9 *Field studies of hillslope flow processes illustrating (a) bedrock microtopographic variation (Brammer and McDonnell 1996); and (b) two-part flow system of deep saturated flow and unsaturated percolation (after Anderson et al 1997)*

percolation of water through the vadose zone colluvium. At the slope base these two flow routes combine in a subsurface saturated zone. Water flowing through fractured rock from upslope forces water percolating through the soil back to the surface, accompanied by mixing of new and old water. The authors argue that under low flow conditions the bedrock flow path dominates, and it is only when the slope base subsurface saturated zone increases in size that significant contributions are made from the vadose zone.

For geomorphologists, two issues of concern therefore are increasing complexity within the subsurface zone and changing topography. The first of these involves process complexity, manifest in detailed field monitoring, while the second has been incorporated to some extent in existing geomorphological models. However, these models stress the fact that topographic change occurs as a consequence of sediment redistribution in the long term. They fail to include processes which link sediment redistribution to topography. One possible advance in this respect, discussed by Beven (1997), involves development of a dynamic 'α' version of TOPMODEL. 'α' represents upslope contributing area, which determines the delivery rate of water from areas further up in the catchment. Expressed as a ratio to local slope angle, this index provides a measure of how much water is likely to be retained at any point in the catchment. 'α' currently maintains a fixed value for each grid element within TOPMODEL through the simulation. Improved model structure and process representation would be gained by developing a version of TOPMODEL to include temporal variation in the upslope contributing area in response to water-table dynamics. Hence the feedback between topography and sediment redistribution could be handled in geomorphological models in this way.

Given the need for geomorphologists to use models for scientific explanation rather than for prediction, in particular, for the solution of the inverse problem, the above examples sound a note of caution. Good models replicate the behaviour of the system sufficiently well that inferences can be made about the processes leading to particular outcomes. It is essential to be fully aware of model limitations in this respect. Process representation requires careful consideration, as most models make *a priori* assumptions about the key processes which need to be included. If these assumptions are incorrect then validation exercises may become meaningless, and validation cannot be used to differentiate between good and bad model formulations. In searching for explanations of system behaviour, these can only be based upon processes which are actually included, and the omission of key underlying processes will bias those explanations.

10.3.3 Scales Appropriate for Hydrological Modelling

A third major issue for geomorphological modelling is that of scale. With regard to hydrological modelling, the issue that has become dominant concerns the relative merits of lumped versus distributed models. Space- and time-averaging is computationally efficient, but has traditionally been assumed to produce models which have poor powers for explaining system behaviour. However, this view has recently been questioned (Van der Perk 1997), through comparison of a variety of models of different complexity and degrees of distribution. With respect to the inverse problem, it is widely accepted that increasing the distributed nature of the system will reduce system modelling error but increase error associated with parameter uncertainty (Yeh 1986). For geomorphological modelling, where spatial and temporal scales could involve almost an infinite number of parameters, establishing an appropriate scale for analysis is central. Guidance can be found in recent research which estimates entropy in infiltration parameters for various degrees of resolution (Figure 10.10). The implication of this is that there is a representative elementary area for optimum model parameterisation, this being of the order of 1 km². Given this suggestion, two recent technological advances offer possibilities for building geomorphological models. Slight modifications are required, since one advance, that of digital terrain modelling, deals with levels of resolution somewhat smaller than 1 km², while the other advance is in general circulation modelling (GCM) and applies to a slightly larger scale. These are discussed briefly below.

10.3.3.1   Small-Scale Resolution within DTMs

Physical process modelling for the geomorphologist has increased potential through developments in digital terrain modelling which enable detailed representation of catchment topography. Water flow pathways are often, although not always, driven by topographic differences (see above) and this has formed the basis for many developments in catchment response modelling. Probably the best example is TOPMODEL (Beven and Kirkby 1979), which has at its heart a topographic index (ratio of 'α' to local slope angle) which drives water flow, and whose range of application has benefited considerably from rapid and detailed topographic digitising techniques.

An issue raised by this advance is the dependence of model output on the resolution of the grid mesh. The example of Saulnier et al (1997) shows clearly how the dynamics of the saturated area of the catchment relates very closely to the topographic index used in TOPMODEL which, in turn, is derived from the DTM. Different scales of resolution, ranging from a grid size of 20 to 120 m, produce different values for the average topographic index and also for the extent of the saturated area (Figure 10.11a). It is also pointed out that it is not just the average topographic index which changes with grid resolution, but also the overall shape of the distribution of topographic index values. Since it is the topographic index distribution within the catchment that determines which regions become saturated, this has implications for predictions of the extent of the saturated area. This example illustrates how these values can be compensated for by adjusting the saturated hydraulic conductivity value derived at the calibration stage (Figure 10.11b). This raises again the question of interpreting system behaviour, given that their range of input values for saturated hydraulic conductivity spans three orders of magnitude. For model calibration a useful exercise would be to investigate whether values of saturated hydraulic conductivity necessary to replicate system behaviour for a given grid mesh size are consistent with field effective parameter values measured *at the same scale*.

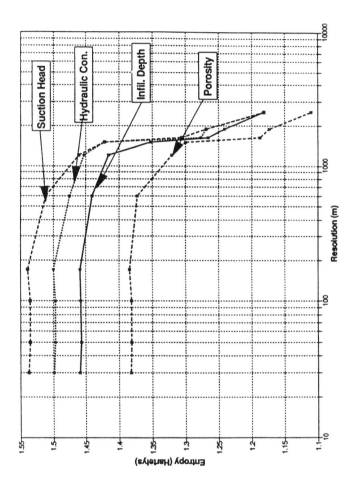

Figure 10.10 The representative elementary area for scales of resolution suitable for geomorphological models (Saulnier et al 1997)

(a)

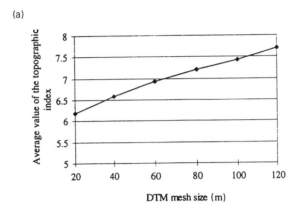

(b)

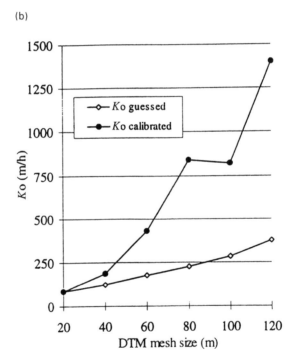

Figure 10.11   (a) Effect of changing DTM mesh size on average values for the topographic index, α, in TOPMODEL. (b) Effect of grid mesh size on calibrated values of saturated hydraulic conductivity (after Saulnier et al 1997)

The integration of DTMs with hydrological modelling has further highlighted the need for caution when trying to interpret the outputs from 'physically based' or 'quasi physically based' models to infer process mechanisms. Such developments also raise the question of whether our modelling goals are predictive or explanatory, and at what scale, in that choice of particular parameter values or model structures may not be physically realistic but may still provide good predictive capability (Van der Perk 1997). Changing the grid size (or scale) greatly influences the way in which different processes are represented, field measurement is carried out and model performance is interpreted.

10.3.3.2   Large-Scale Resolution within GCMs

Climate change is central to geomorphological investigations and, for large spatial scales, GCMs can provide useful estimates. Predictions from GCMs show how variation in the greenhouse gas content of the atmosphere affects global climate (Wigley et al 1990), but, equally, investigations of changes in greenhouse gas content in the past, coupled with reliable GCMs can provide important information concerning past climates. This is useful on a broad scale but, for the geomorphologist interested in individual catchment responses and behaviour in the past, it is critical to know how these global-scale changes affect local meteorological processes. The problem is summarised in Figure 10.12, where the need for downscaling from GCM grid to hydrological model grid is apparent. There is increasing potential to achieve the desired scale of resolution in the data through a variety of downscaling techniques. These are reviewed by Wilby and Wigley (1997) and include regression methods, weather-pattern (circulation)-based approaches, stochastic weather generators, and limited-area climate models.

Several aspects remain for future research on downscaling techniques (Wilby and Wigley 1997). First, downscaling needs to include high altitudes and low latitudes and to consider downscaling for frozen precipitation. Second, weather classifications need to be standardised for larger scales than used at present. Third, downscaling needs to capture spatial and temporal variability in climate at all scales and for all components of the meteorological system. In one example, it has been shown that variation in mean daily precipitation is well represented while interannual variation in annual rainfall totals is not (Conway et al 1996). Finally, there needs to be awareness of the lack of stability in key relationships (e.g. that within a single circulation regime, precipitation diagnostics can vary from year to year).

Integrating across scales raises the fact that outputs from one model can frequently represent the boundary conditions which drive smaller scale models. In using GCM output as input to hydrology models, Hostetler (1994) remarks that it is at the interface of climatic and hydrological models that detailed knowledge of process operation is most uncertain. In particular, modelling processes associated with the ecosphere are especially problematic. If these challenges can be addressed then there is considerable potential for combining GCM output with catchment-scale hydrology models in regional assessments of the impact of

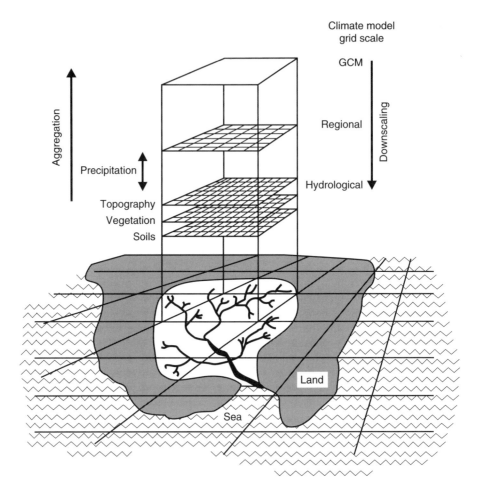

Figure 10.12 *Scales of resolution of different model formulations, illustrating the need for GCM downscaling in geomorphological modelling (Wilby and Wrigley 1997)*

long-term climate change on catchment evolution, a task that lies at the heart of geomorphological modelling.

Both in terms of long-term and short-term physical process modelling, the geomorphologist is faced with the challenge of having to integrate across several disparate scales. On the one hand, upscaling of detailed process investigations can provide a fruitful line of enquiry for understanding the broader scale operation of geomorphological processes, while, on the other hand, downscaling outputs from large-scale models (e.g. GCMs) is central for a more detailed assessment of changing system (e.g. catchment) response under differing boundary conditions (e.g. climatic change scenarios). As Chappell et al (1997) point out with respect to changing scales for permeability characterisation, the former aspect has received far less attention than the latter, but each provides exciting new challenges for physical process modelling at scales appropriate for geomorphological enquiry.

10.4 Lessons from Hydrological Modelling for Future Model Applications in Geomorphology

Physical process modelling encompasses many foci, including atmospheric processes (Cubasch et al 1992; Hewitson and Crane 1992; Hostetler 1994), Earth surface boundary transfers of mass and energy (Jolley and Wheater 1993; Sellers et al 1986), vegetation–soil processes (Wesseling 1998; Wigmosta et al 1994), sediment transport and sediment depositional systems (fluvial, aeolian and glacial) (Fernandes and Dietrich 1997), and the modelling of tectonic responses (Tucker and Slingerland 1994; 1996). Recent advances have expanded the scope of application of physical process modelling to involve coupling with socioeconomic factors such as in the MEDALUS project and as encapsulated in Integrated Assessment Models (Yates 1997), which provide a cost–benefit analysis (economic assessment) as well as considering the physical processes involved. All aspects have relevance for the geomorphologist who is interested in a wide variety of temporal and spatial scales of analysis.

In particular, the long-term concerns of geomorphological modellers provide a unique interpretation of geocomputation, not found in other applications. Most geocomputational techniques, associated definitions and applications are founded on current-day issues. For geomorphologists the concern lies with situations in which process structures can change considerably over time, appropriate scales for discretisation may differ from the detailed structures offered by currently available physical process models, and validation opportunities are impossible for the time frames of interest. This chapter has highlighted several areas which require closer attention when geocomputational methods are introduced into long-term modelling.

Two contrasting approaches have characterised geomorphological modelling, one involving simulation of continuous hillslope change under diffusive processes, and the other involving detailed process-based models parameterised for past time periods. It is important to establish the correct balance between modelling detailed process mechanisms and investigating the net behaviour over larger scales. Developments in physical process modelling of possible relevance to the geomorphologist involve catchment-scale models which emphasise hydrological processes, and lessons from these advances have relevance for geomorphological modelling (e.g. Abbott et al 1986; Beven and Kirkby 1979). Rather than predicting long-term drainage basin change, catchment models provide information on short-term drainage basin response to individual rainstorms (or a series of storms over a few seasons). It is not a simple matter to upscale these responses to describe long term system behaviour. In the long term, the treatment of physical systems requires a qualitatively and quantitatively different approach. Issues which must be addressed in this respect include those of validation, process representation and scale-linkage which currently emerge from applications hydrological models.

To develop geomorphological models further there needs to be a closer focus on the solution of the inverse problem. As geomorphologists attempt to include this to ascertain parameter sets which give rise to particular process outcomes, so model structure and handling of detailed process mechanisms becomes critical. Geomorphological models also require scale-linkage, whereby models which are relevant to different scales are built into a scheme for elucidating long-term, large-scale system behaviour. Hence the output from one model becomes the input to another, requiring that some compatibility exists between models. Using the above discussion we could suggest that the output from a detailed soil hydrology–slope stability model can provide process dominance for development of a diffusivity model for long-term slope adjustment. Similarly, output from GCMs for large spatial scales might be downscaled to provide inputs to catchment models for assessing the effects of climate change on landscape evolution. However, we suggest that it is at scales coarser than 1 km^2 that the most fruitful advances may be possible for long-term applications, despite the fact that this scale involves integration of processes where the detailed mechanics are lost.

Most physically based models used in geomorphology do not adequately represent the processes which govern system behaviour. This has been demonstrated in several recent papers concerned with field monitoring of water flow pathways, the incorporation of vegetation into slope stability models and the representation of upslope contributing areas in catchment hydrology modelling. Related to the scale issue raised above, it may not be detailed hydrology that is required, but better representation of topographic change, involving sediment mobilisation and redeposition, the consequences for hydrological behaviour involving feedbacks with topography and, finally, the inclusion of isostatic readjustment and its relationship with sediment redistribution.

Given the above issues of scale and system representation, including morphology and physical processes, validation exercises are fraught with difficulty. Several new strategies have been suggested in recent research, but there still remains the question of validating models for past time periods or for future scenarios which we wish to understand and explain. There needs to be a change of emphasis in geomorphological modelling towards validation exercises with greater emphasis placed on model structural limitations and the scale required for appropriate parameterisation.

There is currently a very large range of 'off-the-shelf' models available to geomorphologists, many of which run on relatively low-powered machines (Anderson et al 1996). Beven (1997) calls for 'thoughtful' application of TOPMODEL in future applications. However, the general modelling strategy adopted by geomorphologists needs to be 'thoughtfully' applied in the light of issues raised above. These centre around solution to the inverse problem, appropriate validation strategies, process representation and appropriate scales of analysis in the long term. 'Off-the-shelf' models can considerably inhibit the flexibility to address such issues in research. Awareness of such constraints is critical to enable the interface between field or laboratory monitoring and model advances to be used to optimum effect by geomorphologists in a common goal of explaining and understanding Earth sediment transfer systems.

References

Abbott M B, Bathurst J C, Cunge J A, O'Connell P E, Rasmussen J 1986 An introduction to the European Hydrological System – Système Hydrologique Européen 'SHE'. *Journal of Hydrology* 87: 3–59

Anderson M G, Brooks S M 1996 Hillslope processes: research prospects. In Anderson M G, Brooks S M (eds) *Advances in hillslope processes*. Chichester, Wiley: 5–32

Anderson M G, Howes S 1986 Hillslope hydrology models for forecasting in ungauged catchments. In Abrahams A D (ed) *Hillslope processes*. London, Allen & Unwin

Anderson M G, Kemp M, Lloyd D M 1988 Applications of soil water finite-difference models to slope stability problems. *Proceedings of the 5th International Landslide Symposium*, Lausanne: 525–30

Anderson M G, Collison A J C, Hartshorne J, Lloyd D M, Park A 1996 Developments in slope hydrology–stability modelling for tropical slopes. In Anderson M G, Brooks S M (eds) *Advances in hillslope processes*. Chichester, Wiley: 799–822

Anderson S P, Dietrich W E, Montgomery D R, Torres R, Conrad M E, Loague K 1997 Subsurface flow paths in a steep unchannelled catchment. *Water Resources Research* 33: 2637–53

Armstrong A C 1987 Slopes, boundary conditions and the development of convexo-concave forms – some numerical experiments. *Earth Surface Processes and Landforms* 12: 17–30

Bates P D, Horritt M S, Smith C N, Mason D 1997 Integrating remote sensing observations of flood hydrology and hydraulic modelling. *Hydrological Processes* 11: 1777–95

Beven K J 1989 Changing ideas in hydrology: the case of physically based distributed models. *Journal of Hydrology* 105: 157–72

Beven K J 1993 Prophecy, reality and uncertainty in distributed hydrological modelling. *Advances in Water Resources* 16: 41–51

Beven K J 1997 TOPMODEL: a critique. *Hydrological Processes* 11: 1069–85

Beven K J, Kirkby M J 1979 A physically based variable contributing area model of catchment hydrology. *Hydrological Sciences Bulletin* 24: 43–69

Bockheim J G 1980 Solution and use of chronofunctions in studying soil development. *Geoderma* 24: 71–85

Box G E P, Hunter W G, Hunter J S 1978 *Statistics for experimenters: an introduction to design data analysis and model building*. New York, Wiley

Brammer D D, McDonnell J J 1996 An evolving perceptual model of hillslope flow at the Maimai catchment. In Anderson M G, Brooks S M (eds) *Advances in hillslope processes*. Chichester, Wiley: 35–60

Brazier V, Ballantyne C K 1989 Late Holocene debris cone evolution in Glen Feshie, western Cairngorm Mountains, Scotland. *Transactions of the Royal Society of Edinburgh (Earth Sciences)* 80: 17–24

Brooks S M, Anderson M G 1995 The determination of suction-controlled slope stability in humid temperate environments. *Geografiska Annaler* 77A: 11–22

Brooks S M, Collison A J C 1996 The significance of soil profile differentiation to hydrological response and slope instability: a modelling approach. In Anderson M G and Brooks S M (eds) *Advances in hillslope processes*. Wiley, Chichester: 471–86

Brooks S M, Richards K S 1994 The significance of rainstorm variations to shallow translational hillslope failure. *Earth Surface Processes and Landforms* 19: 85–94

Brooks S M, Richards K S, Anderson M G 1993 Shallow failure mechanisms during the Holocene: utilisation of a coupled soil hydrology–slope stability model. In Thomas D S G, Allison R J (eds) *Landscape sensitivity*. Chichester, Wiley: 149–76

Brooks S M, Anderson M G, Collison A J C 1995 Modelling the role of climate vegetation and pedogenesis in shallow translational hillslope failure. *Earth Surface Processes and Landforms* 20: 231–42

Brooks S M, Anderson M G, Wilkinson P, Ennion T 1998 Developments in soil hydrology–slope stability modelling for thin soils with vegetation. *Palaeoklimat Forschung* 20 (in press)

Carson M A 1975 Threshold and characteristic angles of straight slopes. *Proceedings of the 4th Guelph Symposium on Geomorphology*: 19–34

Carson M A, Petley D J 1970 The existence of threshold slopes in the denudation of the landscape. *Transactions of the Institute of British Geographers* 49: 71–95

Chappell N, Franks S, Larenus J 1998 Multi-scale permeability estimation for a tropical catchment. In Brooks S M, McDonnell R A (eds) *Geocomputation in hydrology and geomorphology*. Chichester, Wiley (in press)

Chorley R J 1996 Foreword. In Anderson M G, Brooks S M (eds) *Advances in hillslope processes*. Chichester, Wiley: 1–2

Chorley R J, Schumm S A, Sugden D E 1984 *Geomorphology*. London, Methuen

Conway D, Wilby R L, Jones P D 1996 Precipitation and air flow indices over the British Isles. *Climate Research Special Issue* 7: 169–83

Cubasch U, Hasslemann K, Hock H, Reimer E M, Mikolajewicz U, Santer B, Susen R 1992 Time-dependent greenhouse warming computations with a coupled ocean–atmospheric model. *Climate Dynamics* 8: 55–69

Culling W E H 1960 Analytical theory of erosion. *Journal of Geology* 68: 336–44

Culling W E H 1963 Soil creep and the development of hillside slopes. *Journal of Geology* 71: 127–61

Culling W E H 1965 Theory of erosion of soil-covered slopes. *Journal of Geology* 73: 230–54

Dalrymple J B, Blong R J, Conacher A J 1968 A hypothetical 9-unit landsurface model. *Zeitschrift für Geomorphologie* 12: 60–76

De Roo A P J 1996 Validation problems of hydrologic and soil-erosion catchment models: examples from a Dutch erosion project. In Anderson M G, Brooks S M (eds) *Advances in hillslope processes*. Chichester, Wiley: 669–84

Fawcett K R, Anderson M G, Bates P D, Jordan J-P 1995 The importance of internal validation in the assessment of physically based distributed models. *Transactions of the Institute of British Geographers* NS20: 248–65

Fernandes N F, Dietrich W E 1997 Hillslope evolution by diffusive processes: the timescales for equilibrium adjustments. *Water Resources Research* 33: 1307–18

Ferraresi M, Todini E, Vignoli R 1996 A solution to the inverse problem in groundwater hydrology based on Kalman filtering. *Journal of Hydrology* 175: 567–82

Grove J M 1972 The incidence of landslides, avalanches and floods in western Norway during the Little Ice Age. *Arctic and Alpine Research* 4: 131–8

Hack J T 1975 Dynamic evolution and landscape evolution. In Melhorn W N, Flemal R C (eds) *Theories of landform development*. New York, SUNY: 87–102

Harden J W 1982 A quantitative index of soil development from field descriptions: examples from a chronosequence in central California. *Geoderma* 28: 1–28

Hewitson B C, Crane R G 1992 Climate downscaling: techniques and application. *Climate Research* 7: 85–95

Hillel D 1980 *Principles of soil physics*. New York: McGraw-Hill

Hostetler S W 1994 Hydrologic and atmospheric models: the continuing problem of discordant scales. *Climatic Change* 27: 345–50

Howard A D 1994 A detachment-limited model of drainage basin evolution. *Water Resources Research* 30: 2261–85

Innes J L 1983 Lichenometric dating of debris flow deposits in the Scottish Highlands. *Earth Surface Processes and Landforms* 8: 579–88

Iverson R M 1986 Dynamics of slow landslides: a theory for time-independent behaviour. In Abrahams S D (ed.) *Hillslope processes*. London, Allen & Unwin

Jenny H 1941 *Factors of soil formation*. New York: McGraw-Hill

Jolley T J, Wheater H S 1993 Macromodelling of the River Severn. In Wilkinson W B (ed.) *Macroscale modelling of the hydrosphere*. IAHS Publication No. 214: 91–101

Kirkby M J 1971 Hillslope process–response models based on the continuity equation. *Institute of British Geographers Special Publication* 3: 15–30

Kuczera G, Raper G P, Brah N S, Jayasuriya M D A 1993 Modelling yield changes following strip thinning of a mountain ash catchment: an exercise in catchment model validation. *Journal of Hydrology* 150: 433–57

Lamb H H 1982 *Climate history and the modern world*. London, Methuen

Lunati I, Bernard D, Ponzini G, Parravicini G 1998 Inverse problem and upscaling: comparison between the DS (Differential System) Method and a classic statistical one. *European Geophysical Society* 16 (Suppl. II), Pg C436

Mellor A 1985 Soil chronosequences in neoglacial moraine ridges: Jostedalsbreen and Jotunheimen, S Norway. In Richards K S, Arnett R R, Ellis S (eds) *Geomorphology and soils*. London, Allen & Unwin: 289–308

Mroczkowski M, Raper G P, Kuczera G 1997 The quest for more powerful validation of conceptual catchment models. *Water Resources Research* 33: 2325–36

Page M J, Trustrum N A 1997 A late Holocene lake sediment record of the eroison response to landuse change in a steepened catchment New Zealand. *Zeitschrift für Geomorphologie* 41: 369–92

Pilgrim D H, Huff D D, Steele T D 1978 A field evalutaion of subsurface and surface runoff. *Journal of Hydrology* 38: 319–41

Pilotti A, Menduni A 1997 *Earth Surface Processes and Landforms* 22: 885–93

Refsgaard R, Knudsen P 1996 Operational validation and intercomparison of different types of hydrological model. *Water Resources Research* 32: 2189–202

Robertson-Rintoul M S E 1986 A quantitative soil stratigraphic approach to the correlation and dating of postglacial river terraces in Glen Feshie, southwest Cairngorms. *Earth Surface Processes and Landforms* 11: 605–17

Saulnier G-M, Obled Ch, Beven K J 1997 Analytical compensation between DTM grid resolution and effective values of saturated hydraulic conductivity within the TOPMODEL framework. *Hydrological Processes* 11: 1331–46

Sellers P J, Mintz Y, Sud Y C, Dalcher A 1986 A simple biosphere model SiB for use with General Circulation Models. *Journal of Atmospheric Science* 43: 505–31

Shakleton N J, Imbrie F R S J, Pisias N G 1988 The evolution of oceanic oxygen-isotope variability in the North Atlantic over the past 3 million years. *Philosophical Transactions of the Royal Society of London B* 318: 679–88

Skempton A W 1964 Longterm stability of clay slopes. *Geotechnique* 14: 77–101

Tucker G E, Slingerland R 1994 Erosional dynamics, flexural isostacy and long-lived escarpments: a numerical modelling study. *Journal of Geophysical Research* 99: 12 299–312 243

Tucker G E, Slingerland R 1997 Drainage basin responses to climate change. *Water Resources Research* 33: 2031–47

Van der Perk M 1997 Effect of model structure on the accuracy and uncertainty of results from water quality models. *Hydrological Processes* 11: 227–39

Wesseling J 1998 SWAP 20: its theory and some applications. In Brooks S M, McDonnell R A (eds) *Geocomputation in hydrology and geomorphology*. Chichester, Wiley (in press)

Wheater H S, Jakeman A J, Beven K J 1993 Progress and directions in rainfall–runoff modelling. In Jakeman A J, Beck M B, McAleer M J (eds) *Modelling environmental systems*. Chichester, Wiley: 101–32

Wigley T M L, Jones P D, Briffa K R, Smith G 1990 Obtaining sub-grid scale information from coarse resolution general circulation model output. *Journal of Geophysical Research* 95: 1943–53

Wigmosta M S, Vail L W, Lettenmaier D P 1994 A distributed hydrology–vegetation model for complex terrain. *Water Resources Research* 30: 1665–79

Wilby R L, Wigley T M L 1997 Downscaling general circulation model output: a review of methods and limitations. *Progress in Physical Geography* 21: 530–48

Wilkinson P, Brooks S M, Anderson M G 1998 Development of a non-linear optimisation method for locating critical non-circular shear surfaces. In Brooks S M, McDonnell R A (eds) *Geocomputation in hydrology and geomorphology*. Chichester, Wiley (in press)

Willgoose G R, Riley S 1998 The long-term stability of engineered landforms of the Ranger Uranium Mine Northern Territory Australia: application of a catchment evolution model. *Earth Surface Processes and Landforms* 23: 237–59

Willgoose G R, Bras R L, Rodriguez-Iturbe I 1991 Results from a new model of river basin evolution. *Earth Surface Processes and Landforms* 16: 237–54

Yates D N 1997 Approaches to continental scale runoff for integrated assessment models. *Journal of Hydrology* 201: 289–310

Yeh W W-G 1986 Review of parameter identification procedures in groundwater hydrology: the inverse problem. *Water Resources Research* 22: 95–108

Young A, Saunders I 1986 Rates of surface processes and denudation. In Abrahams A D (ed.) *Hillslope processes*. London, Allen & Unwin: 3–30

11

On Complex Geographical Space: Computing Frameworks for Spatial Diffusion Processes

Andrew D Cliff and Peter Haggett

Summary

The movement of human geography towards a behavioural postmodernist framework has been accompanied by a parallel downrating of the role of space in geographical analysis. While we are not in a position to debate the general issues involved, we agree that the role of geographical space, if taken in the simple sense of raw distance, has decreased its role as an explanatory factor, not least in our own area of research, viz. the geographical spread of infectious diseases. But this is to take a very simplified and singular view of geographical space. Geographical spaces come in a wide variety of forms, of which conventional Euclidean space is only a single strand. To abandon spatial explanations because this one variety is less powerful than in previous historical periods is, in our view, to throw away the proverbial baby with the bathwater.

Accordingly, in this chapter, we explore a number of different ways in which geographical spaces may be respecified in a non-Euclidean manner to provide new insights into the way infectious diseases diffuse. To provide comparison with known results from the conventional view of space, we draw upon examples for three respiratory diseases – measles, German measles and whooping cough – for the island community of Iceland over the course of the twentieth century. The technique we use to define these alternate views of space is multidimensional scaling. The chapter concludes by assessing the advantages and disadvantages of the approach, and considers the implications for GIS.

Geocomputation: A Primer. Edited by Paul A Longley, Sue M Brooks, Rachael McDonnell and Bill Macmillan.
© 1998 John Wiley & Sons Ltd.

Of meridians, and parallels, man hath weaved out a net, and this net, thrown upon the heavens, and now they are his own.

(John Donne, 1611)

11.1   Introduction: The Nature of Geographical Space

While the understanding of geographical space retains a central place within geocomputational analysis, the recent movement of human geography towards a post-modernist and post-structuralist framework has shifted attention away from the role of space as an explanatory variable (Simonsen 1996). Although we are not in a position to debate the general issues involved in this movement, we agree that geographical space, if taken in the simplest sense of distance alone, has decreased its role as an explanatory factor in our own area of research, viz. the geographical spread of infectious diseases. But to equate spatial factors with raw distance is to take a very simplified and singular view of geographical space. Geographical spaces come in a wide variety of forms, of which conventional Euclidean space (on a plane) or spherical coordinate space (on the global surface) are only two strands.

The space within which infectious diseases spread is now very complex following, for example, individual contact networks at one level (Schaerström 1996) or inter-metropolitan flows at another (Smallman-Raynor et al 1992). In this context, a GIS that relies on conventional 'box space' is increasingly unrealistic. By box space we refer to the network of points and areas bounded by eastings and northings (on a planar surface) or latitude and longitude (on a spherical surface); these provide an inert framework or spatial infrastructure for data recording. An alternative approach which we explore here is to allow the data themselves to force the spatial framework into position. The simplest example of such 'forced space' is provided by isodemographic maps where areas with large populations are distended to occupy large spatial extents on the map and vice versa.

In spatial diffusion analysis, the forcing factors are dyadic in the sense that space is determined by an operator consisting of two vectors (a dyad). Such dyads may consist of separation elements between two locations (e.g. the observed time in days taken for an infection to move between two places) or the behavioural coherence of two locations (e.g. the cross-correlation between disease time series in two places). For two main reasons, mapping such complex spaces is considerably more difficult than in the isodemographic case. First, the number of observations to be mapped goes up as a power function of the number of locations (n locations yield up to n^2 dyads). Second, while the dyads can be readily mapped into a hyperspace (where the number of dimensions match the number of observations), it is considerably more difficult to map them into the 2-dimensional form of the conventional map.

This chapter looks at the ways in which we can compute spatial frameworks which are appropriate to the complex geographical spaces within which spatial diffusion processes operate. In sequence we (a) briefly review the methods involved, (b) illustrate their application to the spatial spread of infectious diseases and (c) consider their relative merits in relation to conventional mapping.

11.2 Computing Complex Spaces

Multidimensional scaling (MDS) refers to a family of statistical methods by which the information contained in a set of data is represented by a set of points in multidimensional space. These points are located in the space in such a way that the geometrical relationships between them reflect empirical relationships in the data. The term *multidimensional* denotes the fact that the space into which the points are fitted may vary from low-order spaces (e.g. 1-dimensional scales or 2-dimensional 'maps') up to higher order spaces (e.g. 3-dimensional 'cubes' or 4- or more-dimensional hyperspaces).

As discussed in Harman (1960), the mathematical groundwork for multidimensional scaling was laid as early as the 1930s by Young and Householder as part of the general development of principal component and factor models. But the classic development of the method came in psychology two decades later in the work of the Gulliksen–Torgerson group at Princeton (Torgerson 1958) and the Shepard–Kruskal group at the Bell Telephone Laboratories (Kruskal 1964a; 1964b; Kruskal and Wish 1978; Shepard 1962). Subsequently, the basic mathematics has attracted renewed interest from Kendall (1971; 1975) and Sibson (1978; 1979).

11.2.1 The MDS Mapping Concept

The essential ideas behind MDS are most readily illustrated using a geographical analogy. When a conventional map of a portion of the Earth's surface is drawn, some distortion results in transforming a curved segment of the globe onto a flat piece of paper. The particular map projection used will determine the nature and extent of the scale or directional distortion introduced. But, subject to these known errors, all the map projections in common use attempt to ensure that the locations of points on the map systematically reflect their relative positions on the globe.

If, however, we use MDS to represent points on the globe, a map is constructed in which the locations of the points on the map correspond not to their (scaled) geographical locations but to their degree of similarity on some variable measured for them. For example, we might map points into a time space; locations separated by short travel times will appear close together in such a space, while centres linked by long travel times will be widely separated. In general terms, the greater the degree of similarity between places on the variable measured, the closer together the places will be in the MDS space. Conversely, points which are dissimilar on the variable will be widely separated in the MDS space, irrespective of their geographical location on the globe.

This basic idea can be extended to the mapping of points from geographical space into a wide range of different spaces which are significant for our understanding of the geographical incidence and spread of disease. So, for example, if we are studying attack rates per unit of population for a disease, MDS will map those areas with similar rates in close proximity even though they may be far removed geographically. The MDS problem is to find a configuration of n points in m-dimensional space such that the interpoint distances in the configuration match the experimental dissimilarities of the *n* objects as accurately as possible. This may be viewed as an issue of statistical fitting.

11.2.2 The Dissimilarity Matrix

The MDS method uses a *dissimilarity matrix*, which defines the degree of correspondence between locations in terms of a variable. For example, suppose that $X(i \times j)$ represents a matrix which records the number of cases of a disease reported from location *i* in month *j*. Then define A by $a_{ij} = 1$ if $x_{ij} > 0$ and $a_{ij} = 0$ otherwise. Let the matrix $S = AA'$. Then S is a symmetric matrix in which the *ik*th element, s_{ik}, is the number of months in which there were reported cases in both areas *i* and *k*. We conventionally set $s_{ii} = 0$. It might be argued that the larger the value of s_{ik}, the greater the similarity between the time series of *i* and *k*. Other measures of similarity might be defined, and we consider several in Section 11.3.

The matrix S is a similarity matrix, reflecting the similarity between locations in terms of their patterns of incidence of a disease. From S, a matrix D (= $\{\delta_{ij}\}$) of dissimilarities ($\delta_{ij} = K - s_{ij}$ for some constant *K*) can be obtained. This matrix of dissimilarities serves as the basis for MDS. The dissimilarities are fixed given quantities and we wish to find the *m*-dimensional configuration whose distances 'fit them best'. The final locations of the points in the configuration are selected to preserve the rank ordering of the relative distances of the experimental dissimilarities.

The final distance metric may be regarded as a monotone transformation of the rank ordering. By adopting as our central goal the requirement of a monotonic relationship between the observed dissimilarities and the distances in the configuration, the accuracy of a proposed solution can be judged by the degree to which this condition is approached. For a proposed configuration, we perform a monotonic regression of distance upon dissimilarity and use the residual sum of squares, suitably normalised, as a quantitative measure of fit, known as the *stress*. We seek to minimise stress.

If we are to move from statistical fitting to an understanding of the epidemiological implications of the fitted configuration, then the interpretability of the coordinates of the *n* points in the *m*-dimensional space becomes important. By construction, stress will fall as *m* increases, but comprehending empirically point locations in general hyperspaces as opposed to spaces with $m \leq 3$ is difficult. So the empirical analyses in this paper are carried out in two dimensions ($m = 2$). From the latter, we will obtain an MDS map upon which the relative locations of points defined in terms of their similarity may be compared with their locations on a conventional geographical map. Indeed, by starting each MDS run with the geographical areas (= 'individuals') in their natural geographical positions as defined by some Cartesian coordinate system such as latitude and longitude, it is possible to track the progressive adjustment from the geographical map to a best-fit epidemiological space (see Section 11.2.5).

Denote the *n* points of the configuration by $x_i, \ldots, x_n$, and let $x_i = [x_{i1}, x_{i2}]'$, where the second subscript denotes the space dimension. Let d_{ij} be the (Euclidean) distance between the points x_i and x_j. Then we define the stress, S, of the fixed configuration $x_1, \ldots, x_n$ to be

$$
S_{(x_1, \ldots, x_n)} = \min_{\substack{d_{ij} \text{ satisfying } M}} \left[\sum_{i<j} (d_{ij} - \hat{d}_{ij})^2 \middle| \sum_{i<j} d_{ij}^2 \right] \tag{1}
$$

where M is the monotonicity condition.

11.2.3 The Monotonicity Condition

In MDS, two forms of the monotonicity condition are commonly used:

1. *Weak (Kruskal) monotonicity* (Kruskal 1964a; 1964b). Here, the distances fitted in the monotone regression [the $\hat{d}_{ij}$ in equation (1)] meet the condition

$$
\text{whenever } \delta_{ij} < \delta_{rs} \text{ then } \hat{d}_{ij} \leq \hat{d}_{rs}. \tag{2}
$$

   That is, weak monotonicity allows unequal data to be fitted by equal disparities.

2. *Strong (Guttman) monotonicity* (Guttman 1968). In this case the distances are required to be strongly monotone with the data. That is

$$
\text{whenever } \delta_{ij} < \delta_{rs} \text{ then } \hat{d}_{ij} < \hat{d}_{rs}. \tag{3}
$$

Because this criterion does not allow unequal data to be fitted by equal disparities, it almost always results in a final configuration with higher stress than with approach (1). A full comparison of the criteria appears in Lingoes and Roscam (1973). We have used the weak monotonicity condition in the empirical examples described in Section 11.3.

The exact form of M also depends upon how we deal with ties in the dissimilarity values. For *primary* treatment of ties, M is the condition

$$
\text{whenever } \delta_{ij} < \delta_{rs} \text{ then } \hat{d}_{ij} \leq \hat{d}_{rs}. \tag{4}
$$

In this case, when $\delta_{ij} = \delta_{rs}$ no condition is imposed on $\hat{d}_{ij}$, $\hat{d}_{rs}$. For *secondary* treatment of ties, M is the condition

$$
\text{whenever } \delta_{ij} < \delta_{rs} \text{ then } \hat{d}_{ij} \leq \hat{d}_{rs} \tag{5}
$$

$$
\text{and whenever } \delta_{ij} = \delta_{rs} \text{ then } \hat{d}_{ij} = \hat{d}_{rs}.
$$

In practice, secondary treatment of ties is computationally much easier and faster than primary treatment of ties and, for this reason, is used in this chapter. It should be noted, however, that the method employed to deal with ties will affect the shape of the resulting configuration. The stress is invariant under translation and uniform stretching and shrinking of the configuration. We therefore normalise each configuration by first placing its centre of gravity or centroid at the origin and then stretching or shrinking so that the mean squared distance of the points from the origin equals 1.

11.2.4   Computatation

Computationally, MDS consists of two main problems. The first is to find the configuration with minimum stress; this is done iteratively. From a given configuration a new one, with lower stress, is obtained by using the method of steepest descent. Let

$$\nabla\Sigma = \left(\frac{\partial\Sigma}{x_{11}},\dots,\frac{\partial\Sigma}{x_{nm}}\right) \tag{6}$$

denote the gradient vector (evaluated at the current configuration). Then the new configuration has

$$x_{iv}^{\text{new}} = x_{iv} - \phi\,\frac{\partial\Sigma/x_{iv}}{|\nabla\Sigma|}, \qquad i = 1,2,\dots,n;\; v = 1,2,\dots,m. \tag{7}$$

where ϕ is a step size. At each step of the iterative process, the secondary computational problem of monotone regression is tackled – that of finding the values of $\hat{d}_{ij}$ from the fixed known values of d_{ij} of the distances in the current configuration. Kruskal (1964a; 1964b) provides details of a possible algorithm for the determination of the $\hat{d}_{ij}$ and gives suggestions for the step size, ϕ.

11.2.5   The Affine Transformation

We have already commented in Section 11.2.2 that it is possible to seed the MDS computational procedure with an initial configuration consisting of the geographical coordinates of the objects. We may then compare the final MDS solution with this initial configuration and ask how like a conventional geographical map the final solution is. The 2-dimensional correspondence between any initial MDS configuration (such as latitude and longitude), defined by the coordinates (x,y), and the final configuration in an MDS space, specified by the coordinates, (u,v), may be determined using the affine correlation coefficient (see Tobler 1965). This is based upon the affine transformation familiar in GIS. The closer the MDS coordinates are to the true latitudes/longitudes, the higher will be the spatial correlation (and the more 'map-like' the MDS configuration). The beauty of the affine transformation in its full form is that, given two sets of coordinates, the transformation matrix handles all aspects of the transformation (scaling, shear, rotation, translation, projection and overall scaling); see Ahuja and Coons (1968) for details. Thus it is possible to compare the relationship between the same set of objects mapped in two different coordinate systems.

Let x_i, y_i; u_i, v_i denote the rectangular coordinates of the ith point in the two point patterns we wish to compare (the longitudes and latitudes of the centroid of the ith geographical unit and the corresponding coordinates in the MDS space, respectively). In addition, let $W_i = (u_i,v_i)'$ and $Z_i = (x_i,y_i)'$ represent the ith observation pair, $i = 1,2,\ldots,n$, where the primes denote the transpose. The objective is to use least squares methods to estimate the coefficients A and B in the transformation

$$\hat{W}_i = A + BZ_i \tag{8}$$

The regression can be considered as a mapping from the xy plane to the uv plane.

Treating the observations as the components of vectors in this way means the constants become

$$A = (a_1,a_2)' \text{ and } B = \begin{pmatrix} b_{11} & b_{12} \\ b_{21} & b_{22} \end{pmatrix} \tag{9}$$

and the equation to be minimised is

$$\sum_{i=1}^{n} (\hat{W}_i - W_i)\cdot(\hat{W}_i - W_i)' \tag{10}$$

The complete affine transformation is therefore of the form

$$\begin{pmatrix} \hat{u}_j \\ \hat{v}_j \end{pmatrix} = \begin{pmatrix} a_1 \\ a_2 \end{pmatrix} + \begin{pmatrix} b_{11} & b_{12} \\ b_{21} & b_{22} \end{pmatrix} \begin{pmatrix} x_i \\ y_i \end{pmatrix} \tag{11}$$

The *affine correlation*, which is an overall measure of the degree of correlation between the two point patterns, is given by

$$R_{WZ} = \frac{\sigma_u^2(r_{xu}^2 + r_{yu}^2 - 2r_{xu}r_{yu}r_{xy}) + \sigma_v^2(r_{xv}^2 + r_{yv}^2 - 2r_{xv}r_{yv}r_{xy})}{(\sigma_u^2 + \sigma_v^2)(1 - r_{xy}^2)} \tag{12}$$

This is not symmetric and so $R_{WZ} \neq R_{ZW}$.

11.3   Epidemic Spaces: Icelandic Examples

Many quintessentially geographical questions may be posed about disease distributions: How may we detect 'hot spots' of geographically raised incidence of disease? What causes such spatial concentrations? For transmissible diseases, what are the geographical corridors that will be followed as disease spreads from one area to another? How rapidly will the disease move? Are disease patterns and their diffusion tracks constant in time and space? To what extent may results worked out for one disease be applied to another?

In examining such questions, powerful insights can be gained from studies conducted within the disease space appropriate to the questions posed rather than against the conventional geographical map backcloth commonly used in GIS-based studies. MDS provides the methods necessary to construct such arbitrary spaces. In this section, we illustrate some of the spaces that may be employed and the results that can be generated. Monthly reported case data for three respiratory diseases (measles, German measles and whooping cough) present in the North Atlantic island of Iceland are analysed. Using a variety of 'epidemic spaces', we look at the way in which these infectious diseases have reached Iceland and spread among its medical districts. In all examples, the MDS algorithm was initialised with the geographical centroids of the units studied and the affine correlation was calculated between this initial configuration and the final minimum-stress solution. Arising from equation (12), we define $\bar{R} = (R_{WZ} + R_{ZW})/2$ and quote this in figure captions.

11.3.1   Iceland

11.3.1.1   Icelandic Disease Data

The data analysed comprise the monthly reported cases for three infectious diseases (measles, German measles and whooping cough) 1900–90, for each of a consistent set of 47 medical districts whose populations are mapped in Figure 11.1. (Note that the medical district names have slight differences from town names, which accounts for some apparent discrepancies in spelling between different figures and in the text.)

The data are recorded in *Heilbrigðisskýrslur (Public Health in Iceland)*, the annual report of the Director-General of Public Health, and are described in detail in Cliff et al (1981). Between 1900 and 1990, outbreaks of the three diseases occurred in series of distinct epidemic waves, each separated by intervals during which either no cases or only isolated cases were reported. For measles, 95 256 cases were reported in 17 waves; for German measles, 30 455 cases in 15 waves; and, for whooping cough, 51 058 cases in 17 waves. The national time series are illustrated in Figure 11.2, both as reported cases and cases per thousand population. Figure 11.2 shows that, over the course of the century, epidemics of all three diseases became smaller and occurred at more frequent intervals; there has also been a tendency for the gaps between epidemics to become filled with a low-level ripple of cases. As outlined below, this may be attributed to the growth of the susceptible population and to the increasing accessibility of Iceland to virus invasions brought about by transport changes.

One of the crucial features of the Icelandic data that make them so attractive for modelling is that, to support the quantitative record, the chief physician in each of the 47 medical districts was required to submit an annual report to the Chief Health Officer in Reykjavík. An essential feature of these reports was contact tracing

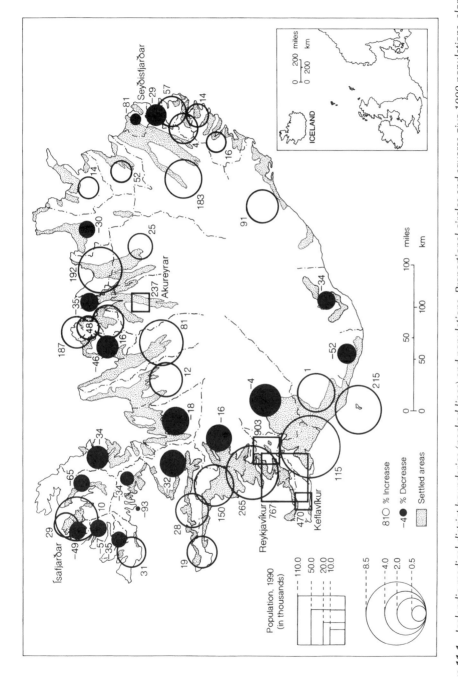

Figure 11.1 Icelandic medical district boundaries (pecked lines) and populations. Proportional circles and squares give 1990 populations, along with population change 1900–90

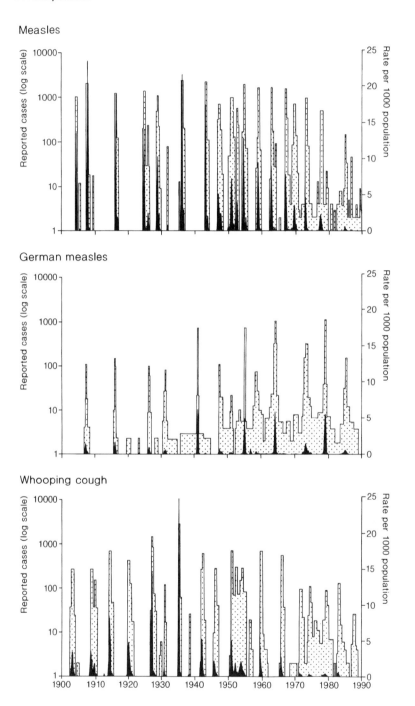

Figure 11.2 *Icelandic epidemics 1900–90. Reported cases of measles, German measles and whooping cough (stippled, log scale) and rate per thousand population (black, arithmetic scale)*

information for each epidemic in each district, giving both the origin of the outbreak and its local spread through the district. Thus, in all essential elements, the pattern of spread of these epidemics through the Icelandic communities is available for comparison with our attempts to identify spread patterns in the data using MDS.

11.3.1.2   Icelandic Demography and Transport

The spread of a disease which relies for its survival upon the mixing of infected individuals (infectives, I) with a large number of potential hosts (susceptibles, S) will be profoundly affected by temporal changes in the demography of the population within which it is circulating and the methods of transport that alter rates of mixing – some of the heterogeneities of Anderson and May (1991: 304–18). Over the course of the twentieth century, Iceland's population has grown from 78 000 in 1900 to 255 000 in 1990. In 1900, 11 per cent of the people lived in the capital, Reykjavík. By 1990, this proportion had increased to 35 per cent, while half of Iceland's population lived in the southwest (Figure 11.1). This regionally differential growth reflects internal migration from rural areas, especially in the northwest and the east, to the Reykjavík area. Away from Reykjavík, settlement is dispersed and coastal (stippled in Figure 11.1).

For all epidemic diseases, there exists a critical community size that must be exceeded for a disease to be endemic in that population. In Iceland, Cliff and Haggett (1990) have estimated that, for measles, this is around 290 000 [cf. Bartlett (1957; 1960), Black (1966) and Schenzle and Dietz (1987) who place the critical community size for measles endemicity between 250 000 and 500 000, depending on geographical isolation]. For German measles and whooping cough, the figures estimated by Cliff and Haggett are 98 000 and 186 000, respectively. So, for measles and whooping cough, no single settlement in Iceland has yet reached the critical community size; for German measles, the Greater Reykjavík region reached the threshold in 1960. Today, Iceland as a whole is approaching the threshold for measles; the country passed the thresholds for German measles and whooping cough in 1925 and 1963, respectively. These facts account for apparent present-day endemicity in the series of Figure 11.2 as compared with the epidemic past.

11.3.1.3   Epidemic Spaces in Iceland

Cliff et al (1981: 50–91) show that a three-stage model is most appropriate for understanding the spatial diffusion of epidemics in Iceland:

1. Initial spread from the rest of the world, generally to the capital, Reykjavík.
2. Spread from Reykjavík to regional centres in the northwest (Ísafjorður), north (Akureyri) and east (Seyðisfjorður/Egilsstaðir).
3. Localised spread from these main centres to their hinterlands.

Table 11.1 *Maps appropriate to different aspects of spatial diffusion processes*

Data type	Conventional maps	MDS maps (2-D)
Point and area data	Populations of medical districts (Fig. 11.1) Medical districts (Fig 11.6c) Temporal changes in MDS distances (Fig. 11.7).	Time–space variations in disease intensity. MDS map based on biproportional scores (Fig. 11.8).
Vector data (Dyadic)	External epidemic pathways (Fig. 11.3a) Internal airline flows (Fig. 11.4a) Internal epidemic corridors (Fig. 11.6a)	External epidemic pathways (Fig. 11.3b) Internal epidemic corridors (Fig. 11.6b).
Similarity data (Dyadic)		Airline connectivity data (Figs 11.4b, 11.4c). Epidemic time-lags contoured onto MDS base (Figs. 11.5a, 11.5b)

Given the generally non-endemic nature of the diseases considered in this paper, for an attack of any of them to occur on a Icelandic farmstead, the causative agent has to cross several hundred miles of sea (or air), travel from the point of landing to a local community, and eventually to the farm itself.

This model suggests three kinds of spaces may be more appropriate for mapping epidemics than a conventional map: first, transport spaces at both the international and national levels to handle the spread between centres which characterises disease diffusion at the macro and meso-scales; second, behavioural spaces to study the mass mixing between susceptibles and infectives essential for the propagation of the disease agents from person to person; third, demographic spaces to study changes over time in the size and frequency of epidemics implied in the concept of critical community size. Table 11.1 illustrates the different kinds of data which are mapped in the eight figures which make up this paper and relates them to both conventional and MDS mapping traditions. We look at each in turn.

11.3.2 An International Transport MDS Space

Figure 11.3a illustrates the known countries of origin for the epidemic waves of one of our target diseases, measles, that affected Iceland over the course of this century. Up until 1943, all introductions are known; after World War II, the record is incomplete. Accordingly, to generate an international transport space, we focused on the period 1900–45, and collected time series giving the annual tonnage of ship arrivals in Iceland from each of the countries shown on Figure 11.3a, including Iceland. Data for the Faroes was amalgamated with Denmark in the official statistics.

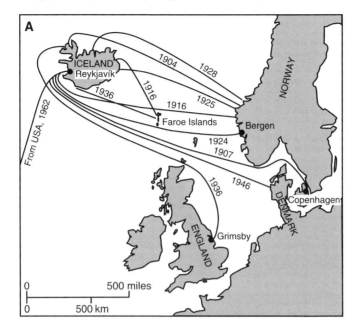

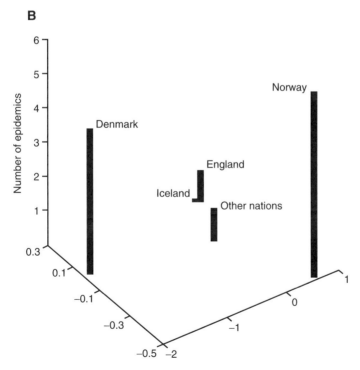

Figure 11.3 *External epidemic pathways. (a) Known countries of origin from Icelandic measles waves 1900–90. Source: Cliff et al (1981, Figure 4.20, p. 90). (b) MDS configuration with number of epidemics started by each country plotted on the vertical axis. R̄ = 0.16*

To generate a similarity matrix for MDS, Pearson's correlation coefficient was calculated between all pairs of time series. The resulting minimum stress configuration in two dimensions is plotted in the x,y plane of Figure 11.3b. The z-axis plots the number of epidemic starts attributable to each country as plotted in Figure 11.3a.

The MDS configuration in Figure 11.3b bears little resemblance to a conventional geographical map, and this is reflected in $\bar{R}$. Denmark/Faroes, with its traditional (former) colonial ties with Iceland, is distinctively isolated from all other countries in the MDS space. Norway is second-furthest from Iceland on the MDS map. The other countries are centrally grouped in the MDS configuration. Just as Denmark and Norway are distinctive in the MDS space, they are distinctive epidemiologically; they were responsible for starting more epidemics of the three target diseases in Iceland than any other country.

11.3.3   An Internal Disease Accessibility Space

Once the disease agent arrives in Iceland, movement of infectives among settlements is required for the generalised dispersal that triggers an island-wide epidemic. If we focus upon the period since 1945, internal air transport has become the main means of transport between the main Icelandic settlements. The rapid growth of internal air travel is shown by the fact that, in 1955, the number of passengers (44 512) represented 1:4 of the then Icelandic population. Today the ratio is less than 1:1. About one-third of the passengers travel on the Reykjavík–Akureyri route (Figure 11.4a). The other main routes, all from Reykjavík, are to Vestmannæyjar (19 per cent), Ísafjorður (13 per cent) and Egilsstaðir (12 per cent) – that is, the main regional centres. The figures for passenger loadings emphasise the hierarchical structure of the airline network illustrated in Figure 11.4a; the main movements emerge like the spokes of a wheel from the Reykjavík hub.

A MDS configuration based upon the airline network was generated from the $(n \times n)$ matrix whose elements were the passenger seats per week on all scheduled internal airline flights between each of the n Icelandic medical districts in summer 1978 (roughly midway through the post-War airline period). Rows represented the origins and columns the destinations to measure potential epidemic 'arrivals'. Similarities were computed using the correlation coefficient.

The resulting MDS configuration is shown in Figure 11.4b. The capital, Reykjavík and the regional centres have been individually labelled. All other medical districts have been coded according to the region of the country in which they are found. Comparison of Figure 11.4b with 11.4a shows that Reykjavík, the hub of the airline network, has been placed centrally in the MDS space. Each of the regional centres (Akureyri in the north, Ísafjörður in the northwest, and Seyðisfjörður/Egilsstaðir in the east) is located peripherally in the MDS space. In addition, Reykjavík, Akureyri and Seyðisfjörður are adjacent to medical districts drawn principally from the regions of which they are the local centres. This MDS space is a classic realisation of the geographical reality of Iceland – the overriding centrality of the capital; the relatively peripheral locations of all other medical districts; and the

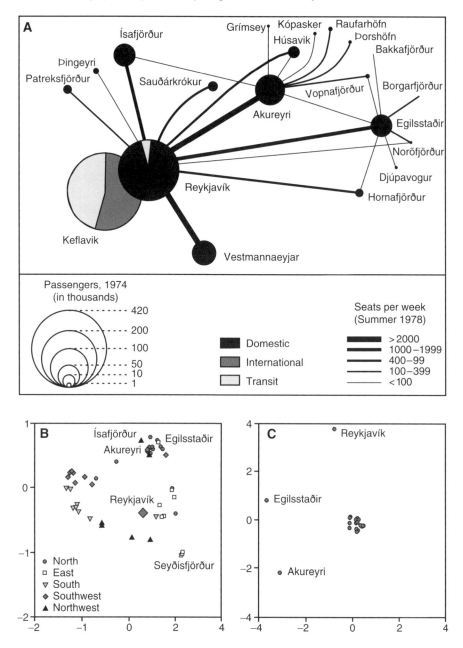

Figure 11.4 *Airline spaces and epidemic velocity. (a) Conventional map of Iceland (coastline omitted for clarity) showing passengers carried 1974 and passenger seats per week on all scheduled internal airline flights, summer 1978. Source: Cliff et al (1981, Figure 5.20, p. 123). MDS representations of airline connectivity are given in (b) using Pearson's correlation coefficient as the measure of similarity ($\bar{R} = 0.56$) and in (c) using a 1/0 coding to denote presence/absence of direct flights to measure similarity ($\bar{R} = 0.09$). In (b) and (c), the capital, Reykjavík, and the main regional centres have been labelled. Other medical districts are coded by their Icelandic region*

dominance of the traditional regional foci over other districts in their geographical hinterlands.

The relative dominance of Reykjavík and two of the regional centres (Akureyri and Egilsstaðir) in the airline system is emphasised in Figure 11.4c. To generate this MDS configuration, the airline matrix was collapsed on a 0/1 basis (0 = no direct passenger flights; 1 = direct flights), and the similarities between medical districts were computed using Tanimoto's dichotomy coefficient, $S6$, defined in Gower (1985). This coding pulls the highly connected nodes of Reykjavík, Akureyri and Egilsstaðir away from all other towns in the least stress configuration.

11.3.3.1   Epidemiological Implications

To investigate the epidemiological implications of Figures 11.4b and 11.4c, the average time taken for epidemics of measles, rubella, and whooping cough to reach each of the Icelandic medical districts 1946–90, was calculated and used to generate a *time-lag surface*. As noted in Section 11.3.1, epidemics of each of the three diseases occurred as discrete events in time, separated by periods during which no cases were reported. For epidemic ℓ of a given disease, code the first month in which a case was reported anywhere in Iceland as month 1 and, for medical district i, note the month in which the disease was reported in that medical district as month 2, or 3, or 4, etc. Denote this month as $t_{i\ell}$. The desired quantities are then

$$\bar{t}_i = (1/k) \sum_{\ell} t_{i\ell} \qquad \ell = 1, 2, \ldots, k, \tag{13}$$

where i is subscripted over the medical districts and k denotes the total number of epidemics of all three diseases observed in the study period.

Figure 11.5 plots the time-lag surface. For each medical district, the average time-lag (in months) to first attack is plotted as the z-axis of a surface in which the MDS airline space of Figure 11.4b forms the base. Note that the z-scale has been reversed, so that surface peaks correspond to short time-lags. Figure 11.5a is a view from the northwest corner of Figure 11.4b; Figure 11.5b is a view from the southeast corner. The surface shows dramatically that the central hub of the airline network, Reykjavík, is attacked on average much earlier than any other Icelandic medical district. As we have noted earlier, Reykjavík is the principal point of international entry of epidemics into Iceland, and it is the epidemic diffusion pole for the rest of the country. Away from Reykjavík, the time-lag surface falls to a series of foothills of long average time-lags in the more inaccessible parts of the country.

11.3.4   Behavioural Spaces

11.3.4.1   Diffusion Corridors

The geographical spread of the respiratory diseases considered in this chapter can only occur by the maintenance of chains of infection as the diseases are passed from

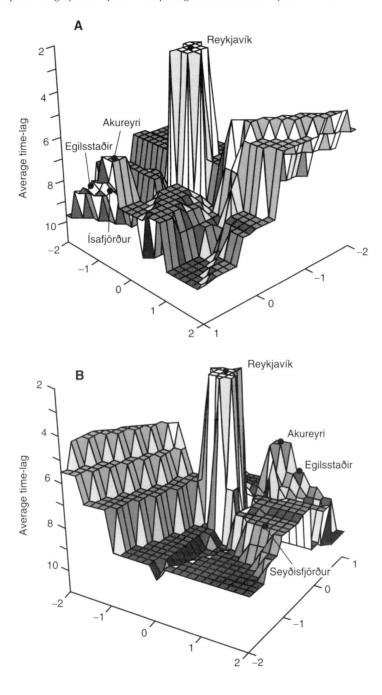

Figure 11.5 *Epidemic time-lag surface. Average time taken in months for epidemics of measles, German measles and whooping cough to reach Icelandic medical districts 1946–90, plotted as a surface over MDS map of airline accessibility. (a) View from northwest corner of Figure 11.4b. (b) View from southeast corner of Figure 11.4b. Note that the peaks correspond to short time lags*

person to person. As noted in Section 11.3.1, the Icelandic records give contact tracing information which indicates, for some of the epidemic waves, the geographical origin of the first case in many of the medical districts affected. Figure 11.6a maps the results of this within-island contact tracing for all epidemics of the three diseases to have reached Iceland this century. Vectors give the number of times these diseases were recorded as propagating between the linked medical districts, and they represent in skeleton form the mixing of susceptibles and infectives that forms the engine of respiratory disease spread. The figure echoes many features of the airline network illustrated in Figure 11.4a. The principal vectors fan out from the medical district containing the capital, Reykjavík, to the three medical districts focused on the regional centres of Akureyi, Ísafjörður and Seyðisfjörður/ Egilsstaðir. Local spread from these centres into their hinterlands is illustrated by the vectors within the stippled areas on each map.

Figure 11.6a was converted into a similarity matrix by assigning rank 1 to the most common epidemic pathway in the diagram (category 6 and over), rank 2 to category 4–5, and so on. Spearman's rank correlation coefficient was then used to calculate the similarity between medical districts in terms of their epidemic pathway ranks. The resulting MDS configuration is illustrated in Figure 11.6b where, echoing Figure 11.5, a contoured time-lag surface has been constructed. Given the uneven scatter of sample points, this should be regarded only as a sketch, but it does indicate the principal links between time-lags to infection and the location of medical districts in the MDS space. Figure 11.6c compares these locations with their true geographical positions in Iceland.

Bearing in mind that the MDS algorithm was initialised with the latitudes and longitudes of medical district centroids, comparison of Figures 11.6b and 11.6c shows how unlike a conventional map of Icelandic medical districts the MDS configuration is. Four main clusters of districts can be identified in the MDS space. As in the airline space of Figure 11.4b, Reykjavíkur is central in the space, confirming its nodality in disease diffusion. In terms of lag-times, comparison of Figure 11.4b with Figure 11.5 again show the increasing time-lags to infection with epidemiological distance from Reykjavíkur.

11.3.5   MDS and Disease Dynamics

So far, we have used MDS to define alternate epidemic spaces. It is equally important to understand the manner in which such spaces may change through time and so, in this section, we indicate how MDS may be deployed to examine the behaviour of disease in time and space.

11.3.5.1   Medical District Dynamics

The qualitative model for disease diffusion in Iceland outlined in Section 11.3.1 raises a number of questions about epidemic behaviour:

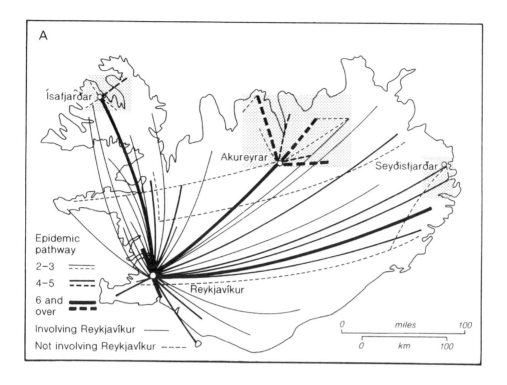

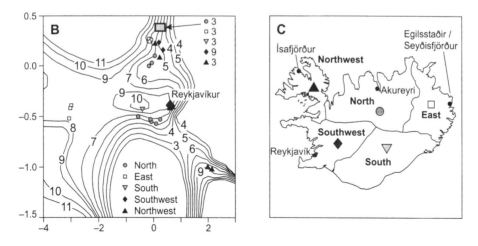

Figure 11.6 *Iceland 1900–90. (a) Observed epidemic corridors for measles, German measles and whooping cough. (b) MDS representation of the epidemic corridors space. The contours show the epidemic time-lag surface constructed for Figure 11.5. (c) Location of medical districts in (b) in conventional geographical space*

1. Have medical districts become more like the capital district, Reykjavíkur, over the course of the century (convergent behaviour) or more dissimilar (divergent behaviour)?
2. Have only the main regional districts of Ísafjarðar, Akureyrar and Seyðisfjarðar/Egilsstaða converged on Reykjavíkur?
3. Have the medical districts in the northwest, north, east and southwest converged on their respective regional districts of Ísafjarðar, Akureyrar, Seyðisfjarðar/Egilsstaða and Reykjavíkur?

Geographically, convergent behaviour implies a 'bolting up' of the spatial structure, whereas divergent behaviour implies that spatial links have weakened and that medical districts are behaving spatially in a more independent fashion.

To examine these questions, the correlation coefficients between the monthly time series of reported cases per 1000 population in each epidemic were calculated and used as measures of similarity in the temporal behaviour of medical districts in that epidemic. MDS configurations were constructed for each of the 49 epidemics of the three diseases that affected Iceland this century, and a number of summary statistics were computed from the resulting configurations. First, the means and standard deviations of the distances in the MDS spaces of each epidemic were calculated between (a) all medical districts and Reykjavíkur (*All* in Figure 11.7), (b) each regional centre and Reykjavíkur (*Ísafjarðar*, *Akureyrar* and *Seyðisfjarðar* in Figure 11.7), and (c) the medical districts in the northwest, north, east and southwest and their respective regional centres (*northwest*, *north*, *east* and *southwest* in Figure 11.7). The means and standard deviations were then plotted as a time series and regression trend lines fitted to check for convergent (falling trend) or divergent (rising trend) behaviour.

Figure 11.7 summarises the results obtained. The left hand semi-circles refer to the means, and the right hand semi-circles to the standard deviations. Circle diameters are proportional to the regression trend line values in 1900 and 1990, and circles have been shaded to denote whether the trend has risen (divergent behaviour, stippled) or fallen (convergent behaviour, black) over the period. If we pool the mean and standard deviation results for each disease then, for measles and whooping cough, 67 per cent of the semi-circles are black; the corresponding figure for German measles is 53 per cent. These findings imply generally convergent behaviour for measles and whooping cough, but much more erratic spatial behaviour for German measles. However, two trends are consistent across all diseases:

1. Means and standard deviations for *All* converge on Reykjavíkur; and
2. The regional centre of *Akureyrar* has converged on Reykjavíkur.

On balance, there is considerable, but by no means unambiguous, evidence to suggest that the epidemic behaviour of Iceland's medical districts for these respiratory diseases has become more similar, and more like that of the capital district, Reykjavíkur as the century has progressed, a trend we would expect as improved

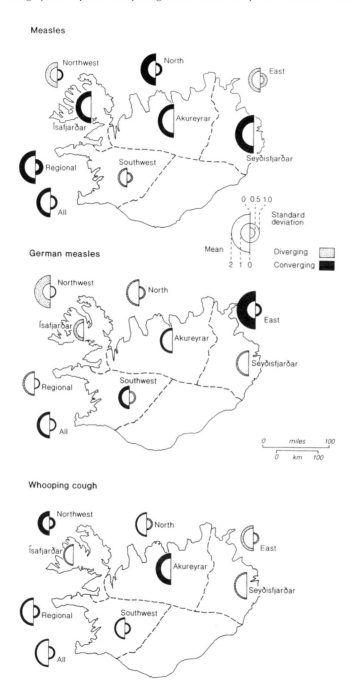

Figure 11.7   *Temporal changes 1900–90, in means and standard deviations of MDS distances for Icelandic medical districts and three infectious diseases. Allowance has been made for epidemic intensity*

internal transport has made all parts of the country more readily accessible to invasions by infectious agents. The trends shown in Figure 11.7 are, however, complex, and the variability exhibited illustrates the changing ways in which space can channel epidemic spread.

11.3.6   Time–Space Changes in Epidemic Intensity

A common epidemiological requirement is to establish whether the rates of infection of any disease vary regionally, and whether any such regional concentrations change over time. One way of examining this question is to take a time–space matrix and to standardise it by computing biproportionate scores. Computational details of the standardisation are given in Bacharach (1970) and Haggett et al (1976: 124–6). Its effect is to generate a matrix of time–space observations in which individual cell values have a score which is greater than unity if the disease is relatively concentrated in that time period and spatial unit when compared with values over the whole time–space system. Scores are less than unity if the disease is relatively unimportant at that time–space coordinate.

Taking reported quarterly case rates for measles in each Icelandic medical district between 1900 and 1990 as an example, biproportionate scores were first computed. Pearson's correlation coefficient was then calculated between the scores for each of the 47 districts to define a measure of similarity between districts on their biproportional scores. This correlation matrix was next used as a basis for generating a 2-dimensional MDS configuration. The resulting plot is shown in Figure 11.8. Selected medical districts have been named, and the remainder have been identified according to the geographical region of Iceland in which they are located. The plot shows very strong spatial clustering of medical districts on a regional basis. The map-like character of the biproportional space is reflected in the average affine correlation, $\bar{R} = 0.65$, the largest affine correlation we obtained. Given that biproportionate scores are a measure of relative time–space intensity, this patterning implies that measles epidemics have been regionally highly distinctive over the twentieth century.

11.4   Discussion

In this section, we draw together the epidemiological findings of our analysis, and then comment upon the utility of the MDS method in defining different epidemiological spaces.

11.4.1   Iceland's Epidemics

Defining different MDS spaces has highlighted a number of features about the twentieth century epidemiological history of Iceland which are referred to time and

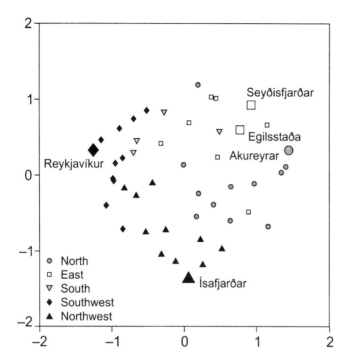

Figure 11.8 *Time–space variations in disease intensity, Iceland 1900–90; an MDS space for measles based upon biproportionate scores ($\bar{R}$ = 0.65)*

again in the physicians' accounts of the epidemics (see Cliff et al 1981: 55–91). In terms of the international origins of many of the epidemics, MDS highlighted the importance of Denmark and Norway as source areas up to 1945 (Section 11.3.2). Once in Iceland, the infective agents for the diseases we considered are diffused from the capital, Reykjavík, to the regional centres of Ísafjörður (northwest Iceland), Akureyri (north) and Seyðisfjörður/Egilsstaða (east); localised spread from each of these centres into their hinterlands also occurs. The MDS spaces defined on the basis of the internal airline network (Section 11.3.3) and epidemic pathways (Section 11.3.4) both emphasised the centrality of Reykjavík as a diffusion pole, and they also picked out the significance of the secondary regional centres in diffusing disease into their local areas.

In Sections 11.3.5 and 11.3.6, we examined time-changes in the diffusion processes. MDS analysis suggested that the epidemiological behaviour of medical districts away from the Reykjavík region tended to become more like that of the capital as the country became increasingly integrated by improvements in transport. It also showed that marked time–space contrasts in epidemic intensity were to be found throughout the century.

11.4.2 Utility of MDS

Like any statistical technique, MDS has characteristic merits and limitations. The signal advantage of the method is that it produces a relatively simple scale-like or map-like output from a very complex input. The degree of fit (stress) between the original structure and the MDS output provides a check on information loss during the summarisation process.

The method offers other advantages for understanding the geographical structure of disease incidence. If we begin the iterative procedure by choosing as starting locations the geographical coordinates of the points in conventional space and calculate a 2-dimensional MDS space, then both the stress and the degree to which the objects move from their seeded positions provide direct measures of how closely epidemic and geographical space coincide. Additionally, the analyses can be carried out at different geographical scales and, by partitioning spatial time series into epidemiologically appropriate time periods, temporal changes in the epidemiological process may be examined.

To set against these merits, the procedure has distinct limitations. We see MDS as akin to the problem of erecting a hideously complex frame tent with sometimes many thousands of tent-poles (dyads), each of different lengths. Although the poles may fit perfectly in a high-enough hyperspace, they become ever more stressed and skewed as the dimensions are progressively reduced towards those lower-order spaces (2-D and 3-D) which the human brain can comprehend. This leads to four problems:

1. *Usability*. With a large number of dyads, the degree of stress in the final 2-D configuration may be too great to provide a useful map. Where this occurs, it is probably best to reduce the number of dyads by decomposing the large mapping problem into a hierarchy of simpler mapping problems at several different levels (e.g. mapping flows between major cities, and then the flows between each major city and the smaller tributary settlements).
2. *Stability*. With a large number of points, the starting positions can modify the final output by trapping points into local optima.
3. *Dimensionality*. Choice of dimensions within which to map the observed similarities is important. Any method in which a multidimensional space is collapsed into fewer dimensions must do an injustice in greater or lesser degree to the true interpoint distances that represent the similarities among the observations in the full space (cf. component and factor analysis which are close relatives of MDS).
4. *Ease of use*. A related drawback is that, like isodemographic maps, the unfamiliarity of MDS maps makes it harder for the reader to retain a sense of his or her bearings. For so many centuries the familiarity of conventional geographical space at local through to global levels has provided a powerful and traditional crutch on which to lean and it will be some time before users become familiar with more challenging MDS space.

But these are still early days for MDS techniques. Cartographers have wrestled over presenting distorted global space on a flat map through many centuries

(Maling 1979): although no absolute solutions have been found, useful compromises (the classic systems of property optimising 'projections') have been worked out. Even at the present rate of progress, it will take many more decades before similar local optima for presenting complex spaces on a flat map will have been found.

11.5 Conclusion

We have argued here that conventional geographical space is only one example of a wide variety of spaces relevant to the study of the spread of communicable diseases. The use of multidimensional scaling as a way of capturing some of these more complex spaces has been illustrated. We also note that, although the discussion here has focused on spatial diffusion processes, the MDS method is capable of being applied to complex mapping problems generally. Haggett (1990: 46–59) has provided a review of geographical applications of MDS ranging from transport problems on Austrian lakes to the reconstruction of Norwegian medieval maps. We suggest that, if progress in spatial modelling and GIS is to flourish, attention needs to be paid to the 'forced spaces' inherent in the process being analysed as well as to the conventional 'box spaces' which dominate most current thinking about spatial analysis.

References

Ahuja D V, Coons S A 1968 Interactive graphics in data processing: geometry for construction and display. *IBM Systems Journal* 7: 188–205

Anderson R M, May R M 1991 *Infectious diseases of humans: dynamics and control*. Oxford, Oxford University Press

Bacharach M 1970 *Biproportional matrices and input–output change*. Cambridge, Cambridge University Press

Bartlett M S 1957 Measles periodicity and community size. *Journal of the Royal Statistical Society A* 120: 48–70

Bartlett M S 1960 The critical community size for measles in the United States. *Journal of the Royal Statistical Society A* 123: 37–44

Black F L 1966 Measles endemicity in insular populations: critical community size and its implications. *Journal of Theoretical Biology* 11: 207–11

Cliff A D, Haggett P 1990 Epidemic control and critical community size: spatial aspects of eliminating communicable diseases in human populations. In Thomas R W (ed.) *London Papers in Regional Science* no 21. London, Pion: 93–110

Cliff A D, Haggett P, Ord J K, Versey G R 1981 *Spatial diffusion: an historical geography of epidemics in an island community*. Cambridge, Cambridge University Press

Gower J C 1985 Measures of similarity dissimilarity and distance. In Kotz S, Johnson N L (eds) *Encyclopaedia of statistical sciences* vol. 5. New York, John Wiley

Guttman L 1968 A general non-metric technique for finding the smallest co-ordinate space for a configuration of points. *Psychometrika* 33: 469–506

Haggett P 1990 *The geographer's art*. Oxford, Blackwell

Haggett P, Cliff A D, Frey A E 1976 *Locational analysis in human geography*, 2nd edition. London, Arnold

Harman H H 1960 *Modern factor analysis*. Chicago, University of Chicago Press

Kendall D G 1971 Construction of maps from odd bits of information. *Nature, London* 231: 158–9

Kendall D G 1975 The recovery of structure from fragmentary information. *Philosophical Transactions of the Royal Society of London A* 279: 547–82

Kruskal J B 1964a Multidimensional scaling by optimizing goodness of fit to a nonmetric hypothesis. *Psychometrika* 29: 1–27

Kruskal J B 1964b Nonmetric multidimensional scaling: a numerical method. *Psychometrika* 29: 115–29

Kruskal J B, Wish M 1978 *Multidimensional scaling*. Beverly Hills, Sage

Lingoes J C, Roskam E E 1973 A mathematical and empirical analysis of two multidimensional scaling algorithms. *Psychometrika* 38: monograph supplement

Maling D H 1979 *Coordinate systems and map projections*. New York, Dekker

Schaerström A 1996 *Pathogenic paths? A time–geographic approach to medical geography*. Lund, Meddelanden från Lunds Universitet Geografiska Institutionen

Schenzle D, Dietz K 1987 Critical population sizes for endemic virus transmission. In Fricke W, Hinz E (eds) *Räumliche Persistenz und Diffusion von Krankheiten*. Heidelberg Geographical Studies 83, University of Heidelberg

Shepard R N 1962 The analysis of proximities: multidimensional scaling with an unknown distance function. *Psychometrika* 27: 125–40 and 219–46

Sibson R 1978 Studies in the robustness of multidimensional scaling: procrustes statistics. *Journal of the Royal Statistical Society B* 40: 234–8

Sibson R 1979 Studies in the robustness of multidimensional scaling: perturbation analysis of classical scaling. *Journal of the Royal Statistical Society B* 41: 217–29

Simonsen K 1996 What kind of space in what kind of social theory? *Progress in Human Geography* 20: 494–512

Smallman-Raynor M, Cliff A D, Haggett P 1992 *International atlas of AIDS*. Oxford, Blackwell

Tobler W 1965 Computation of the correspondence of geographical patterns. *Papers of the Regional Science Association* 15: 131–9

Torgerson W S 1958 *Theory and methods of scaling*. New York, John Wiley

Epilogue

Bill Macmillan

In opening Chapter 2, Helen Couclelis asks what, for this book, must be the key question: is geocomputation to be understood as a new perspective or paradigm or as a grab-bag of useful computer-based tools? Is it, she asks, more than the sum of its parts? In closing the book, I want to ask the question again in a modified form: does the material covered in earlier chapters have sufficient in common to be treated as part of the same enterprise and, if so, what can be said about the enterprise as a whole?

A closely related question has to do with the novelty of geocomputation. Is it any more than the latest round in the use of computers to tackle geographical problems? My answer is 'yes and no'. This is not an attempt to equivocate: *yes* it does include the latest forms of computational geography but *no* this does not mean that it is just an incremental development. It does not consist solely of more efficient computational techniques for solving well-established problems. Technical advances have been so dramatic that the problems we can think about have changed. Moreover, there has been a transformation in the kinds of thoughts we can have. The situation is analogous to the early days of digital computing when the mathematician Richard Bellman said that the very concept of a solution had changed. As it became possible to find solutions to mathematical problems computationally, so it became possible to think about new classes of problems that could be solved. In many cases it would have been fruitless, previously, to contemplate these problems. In others, the problems themselves would have been literally inconceivable.

Couclelis develops her argument by asking a series of related questions about the epistemology of geocomputation and its place both in the discipline and in society at large. One of the central issues is the relationship between geocomputation, quantitative geography and GIS. Couclelis argues that the quantitative revolution

Geocomputation: A Primer. Edited by Paul A Longley, Sue M Brooks, Rachael McDonnell and Bill Macmillan.

'went well beyond the introduction of mathematical and statistical techniques to provide a truly new angle to the conceptualization of geographical space, problems and phenomena'. She makes the perceptive suggestion that if we understand geocomputation as being related to the theory of computation, its impact could be similar to that of the quantitative revolution. This connection with the theory of computation would distance the field from GIS, which is characterised as an 'antagonistic big brother . . . robbing geocomputation of the recognition it deserves'. It would also draw it away from the extreme empiricism with which Openshaw and others would like it to be identified.

We will return to the competing claims of inductive and deductive geocomputation below. Before doing so, something needs to be said about GIS. Goodchild (1992) and others have made the case for the emergence of a geographic information science. In a similar vein, Wright et al (1997) have sought to place GIS on a tool–science continuum. I have argued elsewhere (Macmillan 1998) that this approach confuses categories. We can talk separately about GIS technology, the science of that technology and the scientific work done with the aid of the technology. The argument employs a modification of an analogy drawn from Downs (1997). The analogy suggests that we think of GIS technology as being a 'macroscope' which works by a process of 'minifaction'. Thus, using a GIS is a bit like looking down the wrong end of an optical instrument.

The value of the analogy comes from two observations. First, optical instruments, like the microscope, are, unambiguously artefacts. They are pieces of technology; so are GIS. Second, without contradicting this claim, there are two senses in which optical instruments can be thought of as being scientific: they can be used for scientific purposes, such as the observation of cell structures; and there is a science associated with the instrument itself, namely optics. Making similar distinctions in the context of GIS helps clear the ground for the location of geocomputation. The GIScience advocated by Wright et al (1997) is largely analogous to optics and some of what Couclelis is suggesting might be thought of as falling into this category. But her main concern, and mine, is with the scientific work that can be done with our 'optical' instruments. Such work is, of course, crucially dependent on our equivalent of a sound theory of optics but it is, quite clearly, distinct from it.

The claim I want to stake for geocomputation is that it is concerned with the science of geography in a computationally sophisticated environment. It is also concerned with those computational questions (i.e. questions related to the theory of computation) which are essential for the proper scientific use of our computational instruments. But it is not concerned primarily with GIS technology. To extend the analogy, geocomputation is like astronomy in the post-Galilean world, not like the business of designing and making telescopes. There could not have been a science of astronomy as we know it without the telescope and there are many other things that can be done with telescopes other than astronomical observation but neither of these remarks undermines the position of astronomy as a scientific enterprise.

This brings me back to what was referred to above as the competing claims of inductive and deductive geocomputation. Couclelis is concerned that geocomputation has no philosophy and is proud of it. I would prefer to view the same evidence in a different way. The philosophical position that has been promoted most effectively to date is, essentially, an extreme form of inductivism. In his chapter, Stan Openshaw (this volume) comments on the early life of his Geographical Analysis Machines and refers to 'the tone of GAM', which might be thought of as polemical. It is, more than anything, the tone of Stan that gives geocomputation its inductivist ring. However, in one respect at least, geography is not so very different from astronomy. Of course, observation (and inferences drawn from observations) are vitally important, but so is speculation. Geography would be an odd kind of discipline indeed if it alone eschewed theory and the speculative processes which aid its development.

Having said something about the geocomputational enterprise as a whole we can look at the ways in which the other chapters of this book both contribute to it and amplify our understanding of it. Chapters 3 to 6 of this book are all concerned, in one way or another, with data and their acquisition, retrieval, visualisation and analysis.

Curran et al (this volume) deal with those geographical macroscopes known as remote sensing systems. Their chapter is partly about what we can observe but its main concern is with how we can gain understanding from our observations. It looks at the optical problem of the relationship between image data and the environmental variables that give rise to it. It also looks at the analytical problem of drawing reliable environmental inferences from observational data. From the perspective of the wider geocomputational debate, two of the key issues that arise have to do with data volume and spatial data analysis. Hyperspectral sensors are producing volumes of data that are presenting new challenges for both data handling and visualisation. While the handling problem should be reduced as computing power increases, the visualisation problem is likely to become more acute as remotely sensed data becomes more and more the province of the environmental scientist rather than the specialist in image analysis. Tools are required for visualising n-dimensional datasets and for converting data of high dimensionality into useful environmental information. As the authors note, 'computational methods to extract information from such data are still poorly developed'.

The observations made by Curran et al about the spatial analysis of remotely sensed data are, to the non-specialist, surprising. They note that, although the spatial domain is rich in information, little attention has been given to the development of tools to exploit it. Part of the problem is the nature of the data so the authors go on to identify a variety of spatial statistical approaches that are well suited to remotely sensed data types. They also look at simulated maps, possible realities and fuzzy classification. They do not go on to review neural network techniques and other aspects of artificial intelligence but they say enough to indicate that there is considerable common ground in terms of technical frontiers with other forms of geocomputation.

Goodchild's chapter appears, at first sight, to be quite a contrast to that of Curran et al, but there are important connections. Goodchild is concerned with data of all kinds, including remotely sensed data. And he too can be thought of as dealing with the information produced by geographical macroscopes, not just in the literal sense of Curran et al but also in the figurative sense of population censuses, opinion polls and all the other devices for collecting spatial data. The observation around which his chapter is organised is that 'geocomputation, with its extensive data demands, is arriving as a novel paradigm at a time when many traditional arrangements for production and dissemination of geographic data are breaking down, and are being replaced by a . . . system . . . with far more to offer'.

Building on his experience with the Alexandria Digital Library, Goodchild explores a five-stage process for satisfying a user's data requirements: specify; search; assess; retrieve; and open. Unravelling the complexities of this process requires careful consideration of geographical-information-bearing objects, their footprints and their functional relationships. There are resonances here both with the fuzzy classes discussed by Curran et al and the ideas of Couclelis on the theory of computation and its potential syntactic and semantic contributions to geocomputation.

Goodchild develops the library analogy to good effect but argues that it breaks down at the final stage of opening a dataset for use by an application. He also suggests that the act of selecting only part of a geographical-information-bearing object does not have a paper-library equivalent. Perhaps not, but if we think of the application as an extension of the user – computational reading glasses, as it were – then these two electronic-library activities look rather like old-fashioned book browsing. And this is surely what we are moving towards – having open access shelving rather than stacks from which librarians deliver whole volumes to readers.

Anselin's chapter focuses on exploratory spatial data analysis (ESDA) and visualisation. Like Couclelis, he refers back to the quantitative revolution to make the point that the computational environment has changed qualitatively. Anselin explores the development of ESDA techniques for two data types – geostatistical and lattice. In effect, he takes up the points made by Curran et al about the relative scarcity of spatial statistical techniques for the analysis of certain classes of data and the considerable importance of cartographic visualisation. Thinking of ESDA in terms of presentation, summary and 'potentially explicable patterns' aligns Anselin with Curran et al, Goodchild and Openshaw in that they all seek to provide powerful tools to give ready access to large datasets. In Anselin's case, the tools involve the visualisation of both spatial distributions and spatial associations. His review of dynamically linked spatial association visualisers reveals both the power of the concept and the potential of the new range of closely coupled software systems.

Just as it is surprising for the non-specialist to hear that spatial data analysis is underdeveloped in remote sensing, it is surprising to be told by Anselin that 'the map is conspicuously absent as a separate view of data' in the taxonomy of exploratory data analysis. This presents a considerable challenge which is amplified considerably when space–time data are considered. The vision of animated space–time data explorations is extremely appealing but the computational problems are daunting. As Anselin notes, it is clear that 'methodological developments cannot be considered in isolation from their (geo)computational implications'.

Openshaw's chapter contains the clearest statement in the book of the demand for user-friendly methods for exploratory data analysis: 'We want a push button tool of academic respectability where all the heavy stuff happens behind the scenes but the results cannot be misinterpreted . . . There is also a demand for results expressed as pretty pictures rather than statistics.' What Openshaw offers in response to this demand is the latest in the line of Geographical Analysis Machines (GAMs) – a Geographical Correlates Explanation Machine. The chapter gives a valuable insight into the process by which academic respectability is established. It also answers one of the criticisms of Couclelis: she asserts that geocomputation suffers from, among other things, a lack of major demonstration projects of obvious practical interest. The original GAM work on leukaemia clusters must rate as a demonstration project *par excellence*, of inductive geocomputation at least. On the other hand, it might be said that what is really required is a demonstration that good geocomputational science can lead to effective public action and we are still waiting in the case of leukaemia clusters.

The evolution of analysis machines, running on supercomputers, into what Openshaw calls explanation machines, running on PCs and available over the Web, exemplifies one of the major thrusts of the geocomputational enterprise: to allow an intelligent user to act as a map detective with the aid of a reliable, computational, minifying glass. Openshaw believes, like Sherlock Holmes, that to generate explanations we need no more than good quality data and sound inference. Couclelis begs to differ. The deductive form of geocomputation advocated in her chapter begins to find its voice in the second half of the book but the inductive/deductive partition is, at best, fuzzy. The framework proposed by Couclelis includes, among many other things, experiments with cellular automata. Goodchild suggests that such experiments involve an undifferentiated spatial frame and virtually no data. They can, of course, have these limitations but they don't have to. The essence of deductive geocomputation is that its speculative intellectual exercises are conducted in an environment which is at least potentially data rich. Automata can operate in spatially differentiated, data-filled environments and so can every other geocomputational model entity. Part of the fuzziness associated with the partition has to do with the varying degrees to which different authors bring speculative models and real data together.

The one chapter that does not fit easily into the inductive/deductive partition is Clarke's. This is because it is not primarily about science. It takes Forman's (1987) observation about technology preceding science as the cue for a wide-ranging look at developments in input and output devices and in software. Clarke's thesis, summed up in one of his section headings, is that hardware drives software drives research. I am not so sure about the linearity of this relationship. Going back to Galileo, it is undoubtedly true that telescope technology led the development of optics but it was Galileo's scientific interests that led him to invest time and energy in telescope making. Nevertheless, technological change has a clear impact on the world of ideas.

One of the themes of Clarke's chapter is the increasing portability of computing and the increasing integration of computing devices into everyday life. There is likely to be an increasing fusion of (and, perhaps, confusion between) real and virtual worlds. One manifestation of this process is the possible emergence of a new kind of field computing. He paints a picture of what might be thought of as a human avatar, negotiating through a computationally enhanced real landscape. He describes a Medivac search and rescue by a 'GeoSearch operator' but the image that stays in the mind is pure *Bladerunner*. Clarke is well aware of the dangers of some of the futures he visualises for us. It is salutary to be reminded of them.

The chapter by Batty et al has at least a high fuzzy membership in the deductive class but it also contains plenty of material to connect it with the first half of the book. In particular, it is very much concerned with visualisation, although the realities that are being visualised are largely virtual. One of the key distinctions for Batty et al is that between virtual environments and their users. Another is between real and fictional worlds. The key activities they focus on are representation, modelling and connection. Users can be connected to their virtual worlds in a variety of ways. One of the most intriguing is remote connection over the Web. The authors give a fascinating glimpse of AlphaWorld, a virtual city built by remote 'inhabitants'. They draw from this example a lesson of considerable importance. New urban worlds could be set up for the specific purpose of studying urban growth and form where the worlds rely not on the behavioural rules given to automata but on the decision-making of remote but real users. It has long been taken for granted that the social sciences cannot be experimental because of the impossibility of setting up a social laboratory. Batty et al hint that, in virtual worlds, we will have to rethink what is and what is not possible.

This is by no means all that this chapter has to offer. There are equally instructive forays into the worlds of virtual GIS, virtual modelling and virtual design. Virtual modelling is closely related, in spirit, to the exploratory data analysis ideas of Curran et al, Anselin and Openshaw. Batty et al envisage 'a virtual environment within which mathematical models might be better understood and solutions to problems . . . reached through visualisation'. The specific example given involves optimisation models where a user can drag an icon, representing a facility, across a map and see changes in a spatial-interaction response surface, computed on the fly. This represents, to revisit the Openshaw paper, 'a push button tool of academic respectability where all the heavy stuff happens behind the scenes but the results cannot be misinterpreted'. But there are, in this tool, assumptions about agents and their spatial–economic behaviour, and what is being computed on the fly is what can be deduced about the response of those agents to possible locations of the facility.

Burrough is equally concerned with the business of deduction. His work is cast in what we referred to above as spatially differentiated, data-filled environments and it uses cellular entities with specified properties and relations from which system behaviour can be deduced. Burrough points out the difficulty of linking dynamic models to standard GIS then shows how they can be circumvented using the PCRaster system. The argument is constructed around a set of basic principles for dynamic spatial modelling, predicated on the discrete treatment of space and time. Many of the principles discussed are well established but they appear here in an elaborated and, crucially, fully working form. Moreover, the way they work conforms to Casti's (1997) precepts of good modelling.

Burrough's ideas have wide applicability and very considerable appeal. However, there are important classes of dynamic spatial problem that lie outside his framework. One involves network-based systems, but it is not too difficult to see how a vector equivalent to the PCRaster approach could be crafted. A second involves landscapes that are populated by footloose agents where the properties of both the landscape and the agents may change over time. The challenge which Burrough's chapter poses for social scientists is clear.

Echoing Burrough's concerns about the importance of choosing the correct spatial and temporal resolution for cellular models, Brooks and Anderson give careful consideration to the question of scale. They also look at the problem of integrating across scales and stress both the importance and relative neglect of upscaling. The modelling issues they confront are the fundamental ones of objectives, theoretical content and validation. Surprisingly, these issues do not loom large in other chapters.

Brooks and Anderson argue that it is important to be clear whether a model is to be used for predictive or explanatory purposes because purpose determines form. They also argue that it is vital to get the theory right. It is not sensible to construct a hydrological model based on Darcian flow theory if the model is supposed to represent a catchment where pipeflow is a significant phenomenon. Good model building requires the application of theory that is both sound and appropriate. When it comes to validation, there is, perhaps, scope for a whole book on the form this problem takes in a geocomputational environment but Brooks and Anderson content themselves with the important but counter-intuitive argument that we need to move away from rigorous validation and pay more attention to problems of model structure and scale.

And so to the last chapter on which I will be brief. Cliff and Haggett (this volume) observe that the role of space in geographical analysis has been downrated over the years as 'raw distance' has appeared to decrease in importance. But the study of disease diffusion reveals that it can be highly instructive to employ non-Euclidean geographical forms. Multidimensional scaling assumes the central role in the definition of these alternative perspectives. As in several earlier chapters, interest focuses on dynamic spatial processes. We find that the problems of representation, visualisation and analysis in the context of diffusion are every bit as demanding as those in other geocomputational spheres. And we also find that we can carry forward much of the heritage of spatial analysis in the classical tradition into the geocomputational world.

That world is one in which change is extraordinarily rapid. There is no sense in thinking that it represents a stable new territory which we can map at our leisure. It is changing under our feet. Nevertheless, it has certain major features that are unlikely to be ephemeral. Space and time are generally not separate dimensions. Geocomputation is concerned more than ever before with problems in space–time. And our explorations of such problems continue to emphasise that spatial is special: geographical questions cannot be tackled by the casual extension of techniques from other disciplines. Moreover, in geocomputation, reality is special. It is special in the sense that geocomputation represents a reaffirmation, for anyone who doubted it, that geography is about the real world. But it is also special in that the real world is just one member of the set of all possible worlds. Our ability to explore alternative scenarios and analyse sensitivities is one important element in the geocomputational portfolio. Our ability to visualise patterns and processes in high-dimensional spaces is another. But the key feature of geocomputation as far as I am concerned is the domain it belongs to – the domain of scientific research. Just as astronomy emerged with extraordinary vitality in the post-Galilean world, so geography can emerge from its postmodern slumbers in a geocomputational world. The great attraction of geocomputation is that, while it obliges us to keep our feet planted firmly on the ground, it depends none the less on our willingness to keep our eyes on the stars.

References

Casti J 1997 *Would-be worlds*. New York, John Wiley
Downs R M 1997 The geographic eye: seeing through GIS? *Transactions in GIS* 2(2): 111–21
Forman P 1987 Behind quantum electronics: national security as a basis for physical research in the United States, 1940–1960. *Historical Studies in the Physical and Biological Sciences* 18: 199–229
Macmillan W D 1998 Computing and the Science of Geography. *University of Oxford Economic Geography Research Group Working Paper*, WPG 98-7
Wright D J, Goodchild M F, Proctor J D 1997 GIS: Tool or Science? Demystifying the persistent ambiguity of GIS as 'tool' versus 'science'. *Annals of the Association of American Geographers* 87: 346–62

Author Index

Subject Index